# CLOCKING THE MIND

**Related books**

*Encyclopedia of Social Measurement, Three-volume set, 1–3*
K. KEMPF-LEONARD

*Measurement, Judgement, and Decision Making*
BIRNBAUM

*The Scientific Study of General Intelligence: Tribute to Arthur R. Jensen*
H. NYBORG (ED.)

*The Scientific Study of Human Nature: Tribute to Hans J. Eysenck*
H. NYBORG (ED.)

**Related Journals — Sample copies available online from**
http://www.elsevier.com

*Personality and Individual Differences*
*Intelligence*
*Journal of Personality Research*
*Journal of Mathematical Psychology*
*Organizational Behavior and Human Decision Process*

# CLOCKING THE MIND

## MENTAL CHRONOMETRY
## AND INDIVIDUAL DIFFERENCES

BY

**ARTHUR R. JENSEN**

ELSEVIER

Amsterdam • Boston • Heidelberg • London • New York • Oxford
Paris • San Diego • San Francisco • Singapore • Sydney • Tokyo

Elsevier
The Boulevard, Langford Lane, Kidlington, Oxford OX5 1GB, UK
Radarweg 29, PO Box 211, 1000 AE Amsterdam, The Netherlands

First edition 2006

**British Library Cataloguing in Publication Data**
A catalogue record for this book is available from the British Library

**Library of Congress Cataloging-in-Publication Data**
A catalog record for this book is available from the Library of Congress

ISBN-13: 978-0-08-044939-5
ISBN-10: 0-08-044939-5

For information on all Elsevier publications
visit our website at books.elsevier.com

Printed and bound in The Netherlands

06 07 08 09 10 10 9 8 7 6 5 4 3 2 1

To the memory of
Frans C. Donders (1818–1898)
The Originator of Mental Chronometry

# Contents

# Preface

Mental chronometry is the measurement of cognitive speed. It is the actual time taken to process information of different types and degrees of complexity. The basic measurement is an individual's response time (RT) to a visual or auditory stimulus that calls for a particular response, choice, or decision. The elementary cognitive tasks used in chronometric research are typically very simple, seldom eliciting RTs greater than one or two seconds in the normal population.

Just another methodology for psychology? Mental chronometry undeniably has its own methodology. But it is also much more than just another method of what psychologists know as data analysis. Chronometric methods generate a generically different order of measurement than do any of our psychometric tests.

Scientific research and analysis are rightfully more than just data, numbers, mathematics, and statistical hypothesis testing, yet contemporary psychology has more than its share of these appurtenances. More than in most other scientific fields, research psychologists, especially differential psychologists, confront a plethora of journals, textbooks, and specialized courses on innumerable quantitative methods and statistical techniques. These all are offered wholly without specific reference to any of the field's substantive topics, empirical or theoretical. In the physical and biological sciences typically the methodological aspects and analytic methods are more intrinsically inseparable from the particular phenomena and the theoretical questions that are more or less unique to the special field of study. Why?

A likely explanation of this condition can be stated as a general rule: The lower the grade of measurement used to represent the variables of interest, the more their quantitative description and analysis must depend upon complex statistical methods. Even then, the kinds of questions that can be answered by applying the most sophisticated statistical methods to lower grades of measurement are importantly limited. Quantitative research in virtually all of the behavioral and social sciences is based almost entirely on the lowest grade of measurement that can still qualify as being quantitative, that is, ordinal or rank-order scales. Chronometry, on the other hand, allows us to jump up to the highest grade of measurement, that is, a true ratio scale. Its crucial advantages, though mainly scientific, also have aesthetic appeal. The discovery of the form of a functional relationship between variables when measured on a ratio scale represents a scientific truth, a physical reality, not just an artifice of merely ordinal measurement or of any possible mathematical transformation to which an ordinal scale may be subjected. Measurements can always be transformed from a ratio scale to an ordinal scale, but never the reverse. One derives something

akin to aesthetic pleasure in discovering a fact of nature that is afforded only by true measurement. Many relationships between different behavioral phenomena that we may accept
as true by casual observation or from only ordinal measurement could, in principle, be illu-
sory. Chronometric measurement, however, can in principle confirm or disconfirm beyond
question their validity as scientific fact, of course, always within certain explicit boundary
conditions. Hence, a graphic presentation of chronometric data represents a true relationship, a fact of nature.

An example of this is demonstrated by what I have called the first two laws of individual differences. They just happen to have important implications for interpreting the effects
of education: (1) individual differences in learning and performance *increase* as task complexity increases; (2) individual differences in performance *increase* with continuing practice and experience, unless the task itself imposes an artificially low ceiling on proficiency.
These lawful phenomena are amply demonstrated by a variety of reaction time paradigms.
But they could not be demonstrated definitively with any form of measurement lower than
a ratio scale. Because it has a true or natural zero point, it is the only type of scale that permits fully meaningful comparison of the relative differences between means and standard
deviations, or the variance ratio, $\sigma/\mu$.

It is also of significant importance that chronometric variables are related to psychometric measures of cognitive ability and especially to psychometric *g*. There is also a
*chronometric g* factor, derived from a battery of various chronometric tasks, which is
closely related to psychometric *g*. But chronometric data also reflect a more profoundly
biological basis than do psychometric test scores. For example, it is a fact that IQ, and psychometric *g* even more, measured relatively early in the life span are positively correlated
with individual differences in longevity. It was recently shown in a convincingly large sample that persons' age of death is even more closely related to quite elementary chronometric measures, such as simple and four-choice RT, than it is to IQ (Deary & Der, 2005b). In
fact, the RT measures are the primary basis of the psychometric correlations, as there was
no significant relationship between IQ and longevity when the effect of RT was statistically removed. Also, statistically controlling several social background variables showed
no significant effect on this result.

Individual differences in cognitive abilities must ultimately be explained in terms of the
physical structures and functions of the brain. Much of the research and theorizing in the
field of cognitive psychology has helped in describing more precisely the mental characteristics of variation in psychometric *g* and other psychometric factors than has been possible within the field of psychometrics. But the theoretical formulations of cognitive
psychology are strictly mentalistic, and as such they afford no hooks or handles or lever-
age of any kind for discovering the physical brain processes that accomplish cognitive
behavior or cause individual differences therein. The purely cognitive theories in vogue
today posit an increasing number of metaphorical homunculi each performing some specific functions in a fictional mental realm that has no points of contact with the actual
physical activities of the brain. It is psychology's modern version of Cartesian mind–body
dualism. Individual differences in the speed of information processing, for example, are
either overlooked or attributed to the individual's control of "attention," which along with
"cognitive resources" is one of the hard-worked homunculi in the metaphorical system.
But this approach will not advance us to the promised land, as described by Charles

Spearman in his major work, *The Abilities of Man* (1927), in which he concluded that the final understanding of variation in cognitive abilities "... must come from the most profound and detailed direct study of the human brain in its purely physical and chemical aspects" (p. 403).

At present, explanations of individual differences in terms of actual brain processes are largely confined to studies of the anatomical localization of specific functions. This is now possible by brain-scan techniques, such as functional magnetic resonance imaging (fMRI). It is a most important step in the right direction. As this research advances to more fine-grained analyses of individual differences, it will become increasingly necessary to use simpler but more precise tests of some particular cognitive behavior in order to discover the brain processes involved . This is where chronometry is best suited to make its contribution to brain–behavior research. Chronometry provides the closest noninvasive interface between brain and behavior. Neither the brain nor the behavior can be properly studied in isolation from each other. The common coin for the measurement of both brain activity and overt behavior is real *time* itself, ergo the importance of mental chronometry for research aimed at understanding the brain–behavior relationship. In a few years perhaps enough research of this kind will have been published to warrant a full-scale review, possibly another book.

Before closing this Preface some readers might wish to know of any important published criticisms of my previous work in mental chronometry. The three most competent critiques, as well as the sharpest and the most thoughtful, are provided by Professors John B. Carroll (1987), H. J. Eysenck (1987), and Ian Deary (2000a, 2003).

Finally, I hope that both professionals and graduate students who read this book will readily find many chronometric phenomena that are potential grist for empirically testable hypotheses, and that they will tackle some of the many technical and theoretical problems that are either explicit or implicit in this book.

# Acknowledgments

I am most indebted to the many pioneers of mental chronometry, both past and contemporary who have influenced my thinking and research on this subject. Most are named in the Author Index. I am especially grateful to Professor David Lubinski, the first professional to read my manuscript, for so generously contributing his time and critical acumen in our many discussions of technical issues as viewed by a true blue expert in psychometrics. I am also happy to express my appreciation to Bobbi Morey for her skillful and especially careful proofreading and editing of my original typescript. Whatever defects and oversights remain are regretfully entirely mine.

Every effort has been made to obtain permissions to use figures excerpted from journals and books. We would therefore like to thank the following publishers:

Abelx
Academic Press
American Association for the Advancement of Science
American Psychological Association
American Psychological Society
Blackwell Publishing
Cambridge University Press
Elsevier
Holt, Rinehart, & Winston
Lawrence Erlbaum Associates
Plenum Press
Research Society on Alcoholism
Science
Springer-Verlag
Taylor & Francis
Williams and Wilkins

Chapter 1

# A Brief Chronology of Mental Chronometry

## The Two Disciplines of Scientific Psychology

Psychology as a quantitative, experimental science began with mental chronometry, the empirical study of reaction time (RT). The history of this subject, beginning early in the nineteenth century, is virtually a microcosm of the development of what Lee J. Cronbach (1957) referred to in his famous presidential address to the American Psychological Association as the "two disciplines of scientific psychology" — *differential* and *experimental*. Differential psychology as a quantitative science began with a practical interest in the measurement of individual differences in RT. Experimental psychology, as a discipline distinct from physiology, began with the investigation of the effects of manipulating various external conditions on variation in the measurements of RT. Taken up earnestly in 1861 by Wilhelm Wundt (1832–1920) who founded the first psychological laboratory in Leipzig, RT research became so prominent that the most famous historian of experimental psychology, Edwin G. Boring (1950), wrote that "the late nineteenth century is properly known as the period of *mental chronometry*" (p. 147). To appreciate its significance for psychology's aspiration to become a natural science, we must briefly review some earlier history in philosophy and physiology.

## The Philosophic Background

The work of the German philosopher, Immanuel Kant (1724–1804), was undoubtedly known to virtually all eighteenth century scholars who had an interest in understanding the human mind. Few other philosophers since Plato and Aristotle had a greater influence than Kant in the emergence of psychology as a discipline distinct from philosophy. But Kant's influence was antithetical to psychology's development as a natural science. He was committed to dualism and viewed the mind as beyond the reach of empirical science. He believed that mind had no physically measurable attributes and therefore could not be a subject for research by objective methods. In one of his major works, *Anthroponomy* (1798), he held that even if the expression of mind is mediated by physical structures including the brain, its properties could never be measured because nerve conduction has *infinite* velocity.

Later on, physiologists resisted Kant's dualistic notion and put forth widely ranging conjectures of a finite nerve conduction velocity (NCV). The leading physiologist of his time, Johannes Müller (1801–1858), thought that NCV was 60 times faster than the speed of light. Estimates by other physiologists were much less fantastic. They derived their estimates by incredible arcane reckoning and came up with such amazingly ranging values as 150, 32,400, and 57,600,000,000 ft/s.

## The Introduction of RT Measurement in Physiology

It was not until 1850 that it occurred to anyone to actually measure NCV directly. The physicist and physiologist Hermann von Helmholtz (1821–1894) first did it on bullfrogs, using the nerve-muscle preparation known to all students of physiology, whereby the hind leg's gastrocnemius muscle and its attached nerve are isolated and the nerve is simulated by a fine electrode at various distances from its attachment to the muscle, while the resulting muscle twitch is recorded by a spring kymograph (invented by Helmholtz). The NCV as measured by Helmholtz was between 25.0 and 42.9 m/s. A similar but less invasive method had to be used for humans. Helmholtz measured RT to a cutaneous stimulus. When an electrode was applied to either the toe or the thigh or the jaw, the subject pressed a telegraph key. Thus the distance between these points of stimulation divided by the difference between their respective RTs gave the first realistic measure of peripheral NCV in humans. Values ranged between 50 and 100 m/s, which is comparable to modern measurements. We now know that NCV is not at all a specific value like the speed of light, but varies widely, from about 0.5 to 90 m/s, as a function of certain physical and chemical conditions of the axon. There is a positive relation of NCV to axonal diameter and degree of myelination, variables that vary markedly in different regions of the nervous system.

Boring (1950) assessed this bit of history as follows: "In Helmholtz's experiment lay the preparation for all the later work of experimental psychology on the chronometry of mental acts and reaction times" (p. 42).

## Astronomers' Interest in RT

RT research also had another origin — in astronomy. The story is now legendary. In 1795, England's Astronomer Royal, Nevil Maskelyne, at the Greenwich Observatory, discovered that his assistant, David Kinnebrook, was consistently in "error" in his readings of the time that a given star crossed the hairline in a telescope. He was usually about half a second behind Maskelyne's measurements. This was intolerable for a calibration considered essential for standardizing the accuracy of time measurement for the whole world — the so-called "standard mean Greenwich time." It was Maskelyne's responsibility to ensure the exactitude of its measurement. After repeated warnings to Kinnebrook that he must bring his readings of various stellar transits into agreement with Maskelyne's, Kinnebrook was still unable to correct his average constant "error." So Maskelyne, on the assumption that his own readings were absolutely correct, fired poor Kennebrook, who returned to his former occupation as a schoolmaster (Rowe, 1983). The task that Kinnebrook had " failed" was rather similar to a standard paradigm used in present day RT research, known as "coincidence timing." While observing a computer monitor, the subject depresses a telegraph key as a spot of light begins moving horizontally across the screen and the subject releases the key the instant the moving spot is seen to coincide with a vertical line in the middle of the screen. There are consistent individual differences in the degree of "accuracy" in estimating the time of coincidence of the spot and the line.

The Prussian astronomer F. W. Bessel (1784–1846) read of the Maskelyne–Kinnebrook incident in a report issued from the Greenwich Observatory, and in 1820 he began systematically studying individual differences in the kind of coincidence timing that had led to the firing of Kinnebrook. Bessel indeed found reliable differences between subjects, which he termed the *personal equation.* These measures of individual differences were used to achieve more accurate astronomical timings. The invention of efficient and more accurate electrical and photographic techniques for this purpose came about much later. But over the next 50 years astronomers devised more precise reaction timers, called *chronographs.* The first one was created in 1828 by the German astronomer Respold. A much improved model was produced in 1850 by the United States Coast Survey. Countless individuals' RTs were measured as astronomy journals published data on the personal equation and its practical use in their research, for which metrical precision, then as now, is *sine qua non.* The standard unit of time until recently was the second, defined by astronomers as 1/86400 part of the mean solar day, the time averaged over the year, between successive transits of the center of the sun's disc across the meridian. This definition of the second has proved to be too imprecise for certain researchers in modern physics. So today the most precise standard measurement of the second is rendered by an atomic clock, with an accuracy to within a few billionths of a second per day. The present international standard measure of a second is 9,192,631,770 vibrations of a cesium 133 atom in a vacuum.

The term *reaction time* was coined in 1873 by an Austrian physiologist, Sigmund Exner (1846–1926). He is also credited with discovering the importance of "preparatory set" in the measurement of RT. Lengthening the preparatory interval (i.e., the elapsed time between the "ready" signal and the onset of the reaction stimulus) increases the trial-to-trial variability of the subject's RTs, thereby increasing the measurement error of the mean RT measured in a given number of trials. Since Exner's discovery, the use of a specified or optimal preparatory interval became standard procedure in RT measurement.

## The Experimental Psychology of RT

Research based on RT measurement can be said to have become truly *mental chronometry* in the work of the Dutch physiologist Frans C. Donders (1818–1889), whose research monograph,"On the speed of mental processes," was published in 1868 (for English translation, see Donders, 1969). Donders complicated the RT procedure beyond the measurement of simple reaction time (SRT), or a single response to a single stimulus. He also measured RTs involving *choice* (CRT) and *discrimination* (DRT) between two or more reaction stimuli. By thus increasing the number of different reaction stimuli or the number of response alternatives, or both, he was able to subtract the simpler forms of RT from the more complex to obtain a measure of the time required for the more complex mental operations. For example, sensory and motor components in SRT could be subtracted from CRT, yielding the time taken for the "purely mental" processes involved in making the choice. Donders thought that these experimentally derived time intervals reflected hypothetical mental processes such as discrimination, association, and judgment. Though his RT measurements were accurate, the validity of his theoretical argument based on the *subtraction method* was strongly disputed. It was argued that the total RT is likely not a simple sum of the times taken for each of the hypothesized component processes.

Therefore, subtracting the RT of a simple task (hence a relatively short RT) from the RT of a more complex task (hence a longer RT) leaves a remainder that is not clearly interpretable as the time taken by the hypothesized process that caused the more complex task to have the longer RT. Subsequent research showed that Donders' subtraction method, though clever, was empirically unsupportable. But his contribution remained so influential to subsequent work in mental chronometry that a more detailed discussion must be postponed to Chapter 2, which reviews the various RT paradigms most commonly used in RT research.

Following Donders, experimental research on RT was among the first lines of study pursued assiduously in the world's first psychological laboratory, founded in Leipzig by Wilhelm Wundt in 1879. Throughout the 1880s, RT was the chief activity in Wundt's lab.

One of the notable products of this effort was the doctoral thesis of an American student, James McKeen Cattell (1860–1944), who, with his Ph.D. in experimental psychology, returned to the United States and, after 3 years on the faculty of the University of Pennsylvania, was called to found the psychology department at Columbia University, which he headed from 1891 to 1917. He became one of the most famous psychologists of his time.

Cattell's Ph.D. research, carried out under Wundt's guidance, was later published in the journal *Mind* in 1886, titled "The time taken up by cerebral operations."[1] It began with the ambitious statement: "Mental states correspond to physical changes in the brain. The object of this paper is to inquire into the time needed to bring about changes in the brain, and thus to determine the speed of thought." Much of Cattell's experimental work on RT was strictly parametric, that is, it assessed how RT was affected by the procedural conditions in a given experiment. He discovered that RT was affected by the sensory mode and intensity of the reaction stimulus and by the mode of the subject's motor or vocal response. For example, he invented a lip key and a voice key and discovered that reactions made by the speech organs averaged about 30 ms slower than those made by a movement of the hand. He also found that a subject's RT was affected by varying degrees of attention, fatigue, and practice. All these variables were studied experimentally, so they could be controlled or taken into account in later experiments on the variables of primary interest to Cattell at that time; for instance, the time to distinguish one color from another, which varies for different colors and for color versus its total absence. Similarly, different letters of the alphabet printed even in the same size and style of type differ in discriminability as measured by a subject's RT. The same was found for familiar words. The recognition times for familiar objects are generally longer than recognition times of the printed names of those same objects. But the real significance of Cattell's work on RT lies less in the particulars of his findings than in his guiding philosophy of science, with its emphasis on a strictly naturalistic, objective, and quantitative methodology for experimental and differential psychology. He was contemptuous of the "mentalistic" and "introspectionistic" psychology that prevailed in the early history of psychology, and he seemed to take pride in his "hard science" stance for advancing psychology. Boring (1950, p. 538) quotes Cattell: ". . . most of the work that has been done by me or in my laboratory is nearly as independent of introspection as work in physics or zoology."

## The Differential Psychology of RT

It was the discovery of individual differences in RT that first made it a subject of scientific interest — but in astronomy, not psychology. The experimental psychology of Wundt paid no attention to individual differences, except as a source of "nuisance" variance that had to be minimized or controlled by "averaging out" its effects over a number of individuals. Investigations in Wundt's experimental psychology lab focused entirely on the general and "lawful" effect that some external stimulus variable had on a particular behavioral response. The functional relationship between the stimulus ($S$) and response ($R$) variables, symbolized as, $R = f(S)$, was presumed to apply in some lawful manner to all human beings. In *differential psychology*, on the other hand, consistent differences between individuals in traits and capacities are the main variables of interest. So the above conceptual equation then includes the organism ($O$), i.e., $R = f(S \& O)$. In addition to just two classes of variables ($S$ and $R$) that need to measured, there is a third class of variables — the $O$ variables, or the consistent $R$ differences between organisms all given the same $S$. The need to measure the $O$ variables gave rise to the discipline of psychometrics and the attempt to explain $O$ variation gave rise to the discipline of behavioral genetics, employing the methods of both quantitative and molecular genetics in the causal study of individual differences.

The English scientist and polymath Sir Francis Galton (1822–1911) is generally credited as the founder of differential psychology.[2] He was the first one to use measures of RT, along with a battery of diverse sensory–motor tests, for the express purpose of determining the mean and range of human individual differences in various physical traits and behavioral capacities. His subjects were the London public, or at least those who volunteered to be tested in his Anthropometric Laboratory in London's South Kensington Museum. Subjects had to pay threepence to be tested and twopence to be retested on another day. From 1884 to 1890, Galton with his assistants obtained data on his battery of objective behavioral tests and physical measurements from more than 10,000 men, women, and children (including England's Prime Minister, Sir William Gladstone).

Galton conjectured that individual differences in intelligence, or what he preferred to call general mental ability, would be reflected by measures of sensory discrimination and speed of response to external stimuli. Galton was impressed by the theories of his illustrious older cousin, Charles Darwin, and viewed general mental ability as a product of the evolutionary process. He believed that, in the course of human evolution, fineness of sensory discrimination and quickness of response would have been advantageous in the competition for survival and hence was subject to natural selection. And individual differences in a trait are the essential raw material for the operation of natural selection upon that trait. So he invented various devices for objectively measuring these capacities, including RT to visual and auditory stimuli.[3]

Galton created his own reaction timer, described succinctly in his autobiography (Galton, 1908):

> It was difficult to find a simple machine that would register the length of
> Reaction Time — that is, the interval between a Stimulus and the Response
> to it, say between a sharp sound and the pressure of a responding finger on
> a key. I first used one of Exner's earlier instruments, but it took too much

time, so I subsequently made one with a pendulum. The tap that released
the pendulum from a raised position made the required sound, — otherwise
it made a quiet sight-signal, whichever was wished, — and the responding
finger caused an elastic thread parallel to the pendulum and swinging with
it to be clutched and held fast, in front of a scale, graduated to 1/100ths of
a second. This acted well; there was no jar from seizing the elastic thread,
and the adjustments gave no trouble. (p. 248)

Galton compared the average RT of his subjects grouped by age, sex, and occupation
As intelligence tests had not yet been invented, the only clue to the relationship between
RT and intelligence was the the subject's occupation, classified into seven categories: professional, semiprofessional, merchant-trades person, clerical-semiskilled, unskilled, gentleman, student or scholar. (Galton placed his own test record in the category "gentleman,"
as he was independently wealthy and never held a salaried job.) It must have been something of a disappointment to Galton to find how very little these occupational categories
differed, on average, in visual and auditory RT. Galton wrote virtually nothing about this
correlational aspect of his data and entirely gave up this line of investigation after 1890,
although he was fully occupied with other scientific subjects in the remaining 21 years of
his life. He died just 1 month short of his 90th birthday.

Most psychology textbooks up to the present time have emphasized the seeming failure
of Galton's tests of sensory discrimination and RT to correlate with common sense criteria of intelligence, in contrast to the success of Alfred Binet's (1857–1911) famous test,
published in 1905, which comprised much more complex mental tasks than those used by
Galton and was the progenitor of the Stanford–Binet Test and most later IQ tests.

In retrospect, the explanation for the apparent failure of Galton's tests is unsurprising
simply because of the psychometric shortcomings of his data and the fact that the methods
of statistical inference, particularly R. A. Fisher's analysis of variance, had not yet been
invented. Modern statistical methods would have permitted the discovery of significant
relationships between Galton's tests and other subject characteristics. But psychologists
were too quick and eager to cast a pall over the possible promise of Galton's tests as measures of anything important for understanding the nature of individual differences in human
mental traits.

Fortunately, the original record sheets of nearly all of Galton's massive data were well
preserved by the Galton Laboratory at the University of London, permitting a team of
American psychologists led by Ronald C. Johnson (1985) to study these data by modern
methods of statistical inference. On nearly all of Galton's physical and behavioral measures, including RT, the analysis of variance showed highly significant differences (though
of small effect size) in the theoretically expected direction between the various occupational categories. This modern analysis of Galton's data proves that his hypothesis was
indeed on the right track. It was eventually corroborated in modern times by methodologically rigorous research and is now recognized as of major theoretical importance in the
study of human abilities.

The overall means of Galton's visual and auditory RT data were typical of those found
with modern equipment, indicating that his timing device and procedures yielded reasonably accurate measures — but only *on average* over many subjects. The problem in trying

to use the RT measurements of *individuals* for determining their correlation with other variables, such as occupational level, was the very low reliability of Galton's measurements of an individual's RT. The test–retest correlations (i.e., the reliability coefficient) for both auditory and visual RT were only about .20, because only a single trial was given to each subject. Such low reliability for a single trial, however, is about the same that would be obtained with the best modern RT equipment. Several practice trials followed by 60 test trials are typically required for simple RT to attain a reliability coefficient of about .90. Given the fact that occupational level is typically correlated about .50 with IQ in the general population and the disattenuated correlation between simple RT and IQ is only about .20, it is not surprising that an expected correlation of .10 (i.e., .50 × .20) between RT and occupations would be hardly detectable with a measure of RT that has a reliability of .20. Because measurement error averages out in the mean of a large subject sample, however, one should expect to find mean differences, albeit small, between the occupational groups. Applying the analysis of variance to Galton's RT data in fact shows statistically significant mean differences in RT between the various occupational groups (Johnson et al., 1985).

Galton's tests and the theory behind them lay dormant until James McKeen Cattell became head of psychology at Columbia University in 1891. After receiving his Ph.D. in 1886, he visited England. Though he had received his degree under Wundt, who he greatly esteemed, his reading of Galton's works, such as *Hereditary Genius* (1869) and *Inquiries into Human Faculty* (1883), made him a much more enthusiastic admirer of Galton's thinking than of Wundt's. As Cattell's interest became focused on the nature and measurement of individual differences in human capacities, he especially wished to become personally acquainted with his hero Galton. And indeed Galton's influence was evident throughout Cattell's subsequent career. More than 40 years after his first meeting with Galton, Cattell described him as "the greatest man I have ever met."[3]

With the resources for laboratory research allowed Cattell at Columbia University, he developed an extensive battery of devices for measuring sensory discrimination and RTs similar to Galton's. In fact, Cattell coined the term "mental test." One of his early students, Clark Wissler, used Cattell's Galtonian test battery in the research for his fateful Ph.D. dissertation (Wissler, 1901). The publication of its wholly negative and unpromising conclusions, issuing from the then most prestigious psychological laboratory in America, cast a pall over the further use of Galton's methods in the study of individual differences for more than half a century.

Here was the problem: Wissler's subjects were students in Columbia University, an Ivy League institution with a student body highly selected for academic ability. Most such students are well above the 75th percentile of the IQ distribution in the general population. Hence, Wissler's study sample had a highly restricted "range of talent" in general mental ability, a condition now well-known to underestimate any mental ability correlations that would obtain in the general population. Moreover, because there were no IQ tests at that time, Wissler used students' course grades in classics and mathematics as the criterion measures for intelligence. We now know that in highly selective colleges the correlation between intelligence test scores and college grades is quite low. Worst of all, the reliability of Wissler's RT tests could hardly have been much higher than .40, as his RT measurement for each subject was the average of only three trials. Given these handicapping conditions — the restricted range of talent in the subject sample, the low criterion validity of the intelligence measures, and the low reliability of the RT measures — the correlation between RT and college grades was bound to be exceedingly small. In fact, the Pearsonian correlation was a nonsignificant −.02. A similar fate naturally befell the tests of sensory discrimination as well.

The outcome of Wissler's defective study reinforced the conclusions of an even worse study conducted earlier by Ella Sharp (under the direction of the famous psychologist Edward B. Titchener (1867–1927) at Cornell University. Sharp's (1898–1999) study not only had all the defects of Wissler's study, but was conducted on a sample comprising only seven postgraduate students. Such was the psychometric and statistical naïvete of psychologists at the beginning of the twentieth century.

The historical significance of these studies, however, is certainly not what they found or concluded, but the way they were treated in the American psychological literature. They were repeatedly cited in psychology textbooks as "devastating" evidence against the Galtonian idea that such relatively simple mental processes as sensory discrimination and RTs had any relation to the higher mental abilities or intelligence as reflected by the performance of complex tests requiring reasoning and problem solving. For many years, a majority of psychology books presented these studies as the death knell of Galtonian ideas about mental testing. This erroneous verdict became common knowledge to most psychologists. Deary (1994) provides a vivid account of the biased and misleading treatment of these studies in the psychological literature for more than half a century.

This attitude persisted despite the fact that all the deficiencies (particularly measurement error) and the incorrect conclusions of Wissler's study had been emphatically pointed out early on by Charles Spearman (1863–1945), who later became known as the founder of classical test theory. His path-breaking paper (Spearman, 1904) was published 2 years before he received his Ph.D. in Wundt's department in the University of Leipzig. Spearman not only noted the unwarranted acceptance of the null hypothesis based on Wissler's data but also introduced the mathematical formula for the correction of correlations for attenuation (unreliability), showing that Wissler's data were altogether too unreliable to disprove Galton's theory. Spearman also presented correlational evidence consistent with Galton's hypothesis (Jensen, 2000). In the same paper he also introduced the important mathematical innovation — *factor analysis* — for which he is most famous. Factor analysis is a method for analyzing a whole matrix of all the correlations among a number of different variables to reveal the latent sources of variance that could account for the correlations among many seemingly diverse tests or other variables. For instance, Spearman (1904) showed that a test of pitch discrimination shares a factor (i.e., a latent trait or hypothetical source of variance) that is common to various complex tests of mental abilities and scholastic achievements.

Later but seldom cited studies of RT, are those by Peak and Boring (1926), Lemmon (1927), and Roth (1964). All of these reported significant correlations of RT with standard intelligence test scores (briefly reviewed in Jensen, 1982). More recent historical landmarks in RT research, both in experimental and differential psychology, are presented in later chapters dealing with the specific topics to which they are most relevant.

## Notes

1. Long considered as one of the classics of experimental psychology, Cattell's (1886) published thesis in *Mind* is now available on the Internet: http://psychclassics.yorku/ Cattell/Time/part1-2.htm. A collection of Cattell's writings about his experience with Wundt and other autobiographical material of historic interest has been edited by Sokal (1981).
2. A brief account of Galton's diverse research and his many remarkable achievements, with key references to the literature about Galton, are given in Jensen (1994, 2002).
3. Cattell, J.M. (1929). Psychology in America. *Science*, *70*, 335–347.

# Chapter 2

# Chronometric Terminology and Paradigms

## Basic Terminology

Standardized terminology is an advantage to the exposition and development of any field of knowledge. Mental chronometry has its own terminology, and it will be most efficient if its common elements are all defined in one place rather than having them scattered throughout this book. Although there is no officially standardized form of these terms or their acronyms, certain forms are used more frequently in the literature, and it is these (presented initially in bold face) that will be used consistently throughout this book. Common alternate terms are mentioned in italics. It is understood that such terms and definitions are merely denotative labels. As arbitrary conveniences their only possible virtue is being explicit, clear, and consistent. The terms listed here apply to the generic reaction time paradigm in which a person is presented some kind of stimulus to which some kind of timed response is made.

**Person (P).** The individual being tested. Also called *Subject* (S) or *Observer* (O).

**Experimenter (E).** The individual who administers the test.

**Stimulus (S).** A change (on or off) in some stimulus condition (light, sound, etc.) detectable by a sensory receptor; also *Signal*.

**Reaction (R).** A detectable movement, physical change, or action by P (e.g., pressing or releasing a Morse key or push button) occasioned by the occurrence of S; also *Response*.

**Reaction Stimulus (RS).** A designated S to which P is instructed to react in some specified way (e.g., a light going on or off).

**Preparatory Stimulus (PS).** A stimulus occurring prior to the RS intended as a ready signal or cue to alert P of the impending RS. To avoid confusion between PS and RS, the PS is usually in a different sensory modality than that of the RS; also *Forestimulus*.

**Preparatory Interval (PI).** The interval occurring between the onsets of the PS and the RS; also *Foreperiod* (see Figure 2.1).

**Warning Interval (WI).** The interval between the termination of PI and the onset of RS.

**Reaction Time (RT).** The elapsed time between RS and R, usually measured in milliseconds (ms), when P has been made aware that this is a timed response, which should be made quickly but accurately. Also *Decision Time* (DT), used particularly (but unnecessarily) in paradigms in which P's total response to the RS is divided between the RT and movement time (MT) (see below) (see Figure 2.1).

**Response Time (RT°).** Luce (1986, p. 2) introduced this useful distinction between RT and Response Time (which is here symbolized as RT°), the former referring to testing conditions in which reaction speed is an explicit and major focus in the instructions, the latter to conditions in which timing and speed are either not emphasized or not mentioned in E's instructions to P. (The effect on RT of instructions emphasizing speed versus accuracy is discussed in Chapter 3.)

**Home Key (HK).** Also *Home Button.* In some procedures P presses down a key (or button) while awaiting the onset of the RS. When RS occurs, P releases the key (i.e., lifts finger) and presses another key called the **Choice Key (CK)** located some distance from the HK. RT is the interval between the onset of RS and P's releasing of the HK (see Figure 2.2).

**Movement Time (MT).** The interval between P's releasing the HK and touching the CK.

**Simple Reaction Time (SRT).** Elapsed time between the onset of a single RS (e.g., a light) and a single intentional response (e.g., pressing or releasing a key).

**Choice Reaction Time (CRT).** The time interval for an intentional response to a choice between two (or more) different RS, one of which is designated as the positive RS⁺, the other(s) as the negative RS⁻. Also *Disjunctive Reaction Time* (i.e., RS⁺ or RS⁻).

**Conditional Choice Reaction Time (CCRT).** This is the same CRT except that P's response to the RS is conditional on some additional feature of the discriminative stimulus. For example, say the two keys on the response console are colored red and black. But there are four possible RSs, which appear in a random order: a red circle on either a yellow or blue background, and a black circle on either a yellow or blue background. P must press the red key when the red circle appears on a yellow background but must press the black key

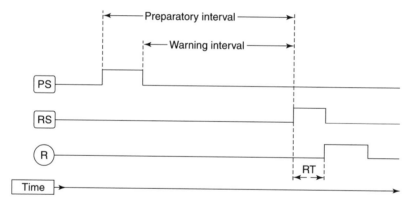

Figure 2.1: The time lines for events in the typical procedure for measuring reaction time. PS, preparatory stimulus; RS, reaction stimulus; R, response.

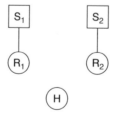

Figure 2.2: A choice reaction time (CRT) paradigm, using a home key (HK). The cycle begins by the person (P) depressing H with the index finger. When $S_1$ or $S_2$ (e.g., a light) appears, P leaves H and presses the appropriate response key ($R_1$ or $R_2$) as quickly as possible.

when the red circle appears on a blue background, and there is a similar cue reversal for the black circle. That is, P's response is simultaneously conditioned on both the foreground stimulus and the background stimulus. (The conditional stimuli can also be presented in different sensory modalities). Having to keep in mind the conditional rule for correct responses strains P's working memory capacity and makes CCRT appreciably longer than CRT; it also increases response errors. Both RT and error rate depend on the degree to which E's instructions emphasize either speed or accuracy of responding.

**Discrimination Reaction Time (DRT).** The same as CRT except that the choice to be made is between the two (or more) RSs that are not clearly and obviously different from one another and hence require P to discriminate between them on a particular sensory or semantic dimension.

**Nondiscriminative Reaction Time (NDRT).** Also *Conjunctive RT.* Two (or more) different RSs to either of which P may respond indiscriminately.

**Practice Trial (PT) and Test Trial (TT).** RT experiments typically consist of a given number of PTs and TTs decided by E. PTs are essential for familiarizing P with the apparatus, the task requirements, and the proper procedure, thereby increasing the consistency and reliability of P's performance.

**Interstimulus Interval (ISI).** In a sequence of RT trials, ISI is the time interval between the onset of the RS on a given trial and the onset of the RS on the next trial; the ISI may be uniform or variable across trials.

**Response–Stimulus Interval (RSI).** In a sequence of trials in an RT experiment, the time interval between the P's response and the onset of the next RS in the sequence; it may be a constant or variable interval as determined by E.

**Reaction Time mean (RTm).** The arithmetic mean $(\overline{X})$ of P's RTs over a given number $(n)$ of trials.

**Reaction Time median (RTmd).** The median of P's RTs over $n$ trials.

**Reaction Time Standard Deviation (RTSD) (also SDRT).** The standard deviation $(SD)$ of P's RTs over $n$ trials — a measure of P's trial-to-trial intraindividual variability or oscillation in RT.

**Latency.** The unobserved or hypothetical time for a reaction to occur, of which the observed RT is an estimate that has some residual error — analogous to the psychometric distinction in classical test theory between an *obtained* score $(X)$ as an estimate of the *true* score $(T)$, where $X$ has a margin of error $(e)$ indicated by the standard error of measurement), hence $X = T + e$.

**"Noise."** Also *Neural Oscillation.* A hypothetical construct accounting for variability in a behavioral response when the external stimulus and conditions are uniform or constant. It is indicated by measures of RT variability, e.g., RTSD.

**Display.** A panel on which the RS is presented, such as by light-emitting diodes (LEDs), pea lamps, or figures, words, etc. projected on a computer monitor or a tachistoscope. For auditory RS (or for a PS) a sound generator, usually capable of producing pure tones of a given pitch, is usually a part of the display unit.

**Response Console.** A panel or mechanism on which P makes responses by means of push button(s) or key(s). The response, usually made with the index finger of P's preferred hand, may be a "press" or a "lift" (i.e., removing the finger to release a depressed key).

**S–R Compatibility** refers to the proximity and the order or pattern congruence between locations of the two or more RS elements and the two or more R elements in a CRT or DRT task. Generally, the higher the degree of S–R compatibility, the shorter is the CRT or DRT.

**Elementary Cognitive Task (ECT).** As defined by J. B. Carroll (1993): "An elementary cognitive task (ECT) is any one of a possibly very large number of tasks in which a person undertakes, or is assigned, a performance for which there is a specifiable class of 'successful' or 'correct' outcomes or end states which are to be attained through a relatively small number of mental processes or operations, and whose successful outcomes depend on the instructions given to, or the sets or plans adopted by, the person" (p. 11). All of the RT paradigms described in the rest of this chapter qualify as an ECT.

## Classical RT Paradigms

In the RT literature, a *paradigm* refers to the particular physical arrangement of the S and R elements of the test, the procedural requirements for performing the task, and the particular time intervals measured as the dependent variable(s) in the test. Any paradigm can be described in purely abstract or generic terms, without reference to details of the particular apparatus or procedures used in a given experiment; and in fact these features of nominally the same paradigm can vary considerably from one laboratory or experiment to another.

The first classification of RT paradigms came from Donders (1868), who labeled them the **a**-, **b**-, and **c**- reactions. Wundt later added the **d**-reaction. Although the Donders–Wundt terminology is seldom used in the present literature, the paradigms themselves are generic and well known.

Donders' **a**-reaction is the time taken to make a single response to a single stimulus, i.e., SRT.

In the **b**-reaction, P is instructed to make different responses to different stimuli, i.e., CRT, therefore requiring a *choice* between stimuli. Its simplest form requires two different RSs and two response keys.

In the **c**-reaction, P is instructed to make a response to a positive reaction stimulus (RS$^+$) and to inhibit response to a negative stimulus (S$^-$). It is the simplest form of DRT, requiring only one response key and two or more discriminative stimuli (RS$^+$ and S$^-$), only one of which elicits a response, hence requiring a discrimination between the positive and negative stimuli.

The **d**-reaction added by Wundt is the same as NDRT in which P is instructed to respond to every one of two or more different stimuli; no choice or discrimination is called for even though various RSs are presented. It requires only one response key and two or more RSs. The **d**-reaction is a necessary addition to Donders' three RT types, because, even though a choice or discrimination is not called for, the **d**-reaction elicits a slightly longer RT than the **a**-reaction, indicating that some discrimination and encoding of the stimuli are still being made, though unintentionally.

### Donders' Subtraction Method

Donders assumed that in any one of the RT paradigms, the measured RT is composed of the time increments required by each of a number of successive *stages*. In each of these stages, a particular process transpires between the onset of the RS and the P's initiation of the response. Donders classified these distinct stages as of two kinds, *physiological* and *cognitive*. In SRT, for instance, first the RS must be registered by a sense organ (*T*1), then is transduced to a neural impulse (*T*2), which then is transmitted via an efferent nerve pathway to a specific location in the somatosensory cerebral cortex (*T*3), where it is neurally analyzed for recognition as the RS (*T*4). The "decision" then is transmitted by an efferent nerve pathway from the brain to the hand (*T*5), where the impulse crosses the myoneural junction (*T*6) which, after a certain muscle lag time (*T*7), causes the arm, hand, and finger muscles to respond with the appropriate movement. The sum of all these time components, *T*1 through *T*7, adds up to the RT, which, for the **a**-reaction (SRT), Donders believed was largely assignable to the *physiological* rather than to the *mental* sources of the RT. (We know now, however, that SRT is actually entirely a physiological phenomenon, which can be tracked through structures in both the peripheral and central nervous systems (CNSs), with most of the time interval constituting SRT attributable to processes in the brain.) Donders, who apparently thought about physiology and mind as would a philosophical dualist, wrote

> The idea occurred to me to interpose into the process of the physiological time some new components of mental action. If I investigated how much this would lengthen the physiological time, this would, I judged, reveal the time required for the interposed term. (Donders, 1868/1969, p. 418)

Hence the **c**-reaction, in which P discriminates between $RS^+$ and $RS^-$ and responds to one but not the other, is an example of an *insertion* of a mental process into the sequence of physiological stages elicited in the **a**-reaction. By subtracting RT**a** from RT**c**, therefore, Donders held that he had measured the time taken for the inserted mental process of discrimination, because all the time components of the **c**-response were presumed to be exactly the same as those in the **a**-response except for the additional time required for the inserted discrimination.

The difference between RT**a** and RT**d** was thought to provide a measure of the time difference between sheer stimulus apprehension (RT**a**) or the simple awareness that a stimulus had occurred without its being explicitly encoded with respect to any specific characteristic (e.g., color, shape, or size) and a specifically encoded stimulus (RT**d**). People tend to be compulsive encoders; they encode stimulus characteristics even when not required to do so, and encoding takes time. Hence RT**d** minus RT**a** yields a positive difference.

Subtracting RT**d** from RT**c** yields a measure of the time required for making an overt *choice* response. All of the different RSs are responded to in the **d**-reaction, but only the $RS^+$ is responded to in the **c**-reaction. Presumably, the only difference between responding to every RS (RT**d**) and having to make the choice of either responding or not responding (i.e., RT**c**) is the time needed for discriminating between $RS^+$ and $RS^-$. So RT**c**−RT**d** equals the time taken for choosing between $RS^+$ and $RS^-$.

Using this subtraction method, Donders claimed to have measured the intervals of time taken up by mental processes such as apprehension, encoding, choice, and discrimination, as well as more complex processes such as association and judgment.

### Objections to Donders' Method of Insertion and Subtraction

The problem with Donders' method is that it was based on an untested theoretical assumption: that the insertion of a particular complicating requirement in a simpler RT paradigm did not affect the other components of RT in the simpler paradigm and that its incremental effect on RT was therefore strictly additive. It assumed that additional processing requirements can be *inserted* into a given task, or can be *deleted*, without in any way affecting the other processes involved in the task. Because Donders' crucial assumption did not stand up to later experimental tests, it became known as the *fallacy* of "pure insertion." The insertions were not always purely or consistently additive in their effects on RT, but interacted with other elements of the paradigm. This made it questionable to conclude that it was possible to measure the time required for the execution of a particular mental process by inserting an additional task requirement. Donders' theory and methods were really too simple to deliver what he had hoped for. His insertion and subtraction method failed to provide the necessary means for testing its essential assumption of the additivity of independent time increments for the successive stages of information processing in various RT paradigms. This does not mean, however, that variation in RT caused by manipulating the RT paradigm in various ways is of no scientific interest. That variation is precisely what continues to dominate experimental work in mental chronometry. The demise of Donders' theory only means that his particular assumptions and limited method of analysis are now recognized as inadequate and must be modified and elaborated by other methods. A suitable method of analysis awaited the invention of analytical and statistical techniques that were nonexistent in Donders' day.

### The Additive Factors Method

This method for improving on what Donders had tried to do was introduced by Saul Sternberg (1969), an experimental psychologist in the Bell Laboratories. His method is essentially based on the mathematical technique known as the *analysis of variance* (ANOVA), invented in the 1930s by the British statistician Sir Ronald Fisher (1890–1962). ANOVA is a rigorous method for designing experiments and statistically evaluating their results. It is especially called for when there are simultaneous *effects* of two or more independent variables (called *factors*) on the dependent variable. (An *independent variable* is one that is controlled by the experimenter, e.g., the particular features of an RT paradigm; a *dependent variable* is the subject's measured response, e.g., RT.) ANOVA divides up the total variance of the dependent variable into independent additive components attributable to each of several sources: (1) the separate *main effects* of the independent (i.e., experimental) variables, (2) the effect due to the *interaction* of the independent variables, and (3) *individual differences* + *measurement error*. (By obtaining repeated measurements with the same subjects, E can get separate estimates of the two components in (3).) In a large experiment, if the interaction component is unambiguously statistically nonsignificant, it can be concluded that the dependent variable (RT in this case) results only from the additive effects of the experimental variables. A statistically significant interaction means that the dependent variable reflects not only the main effects of the independent variables but also an effect arising from some *nonadditive* combination of the separate effects of the independent variables.

What would such an experiment look like in the simplest possible experimental design that could tell us if the different mean RTs resulting from two different experimental effects are purely additive or are interactive? The minimal experimental design needed for this kind of analysis consists of two *factors* (i.e., different RT paradigms) and two *levels* of each factor. Suppose the two RT paradigms (i.e., the *factors* in the experiment) are Donders' **c**- and **d**-reactions. In each paradigm there are two discriminable response stimuli, $RS^+$ and $RS^-$. In the **c**-reaction P responds (e.g., presses a key) when the positive but not the negative RS appears, which requires a discrimination. (In the **d**-reaction, P responds to either RS on every trial, which requires no discrimination, but merely apprehension of the onset of either stimulus.) To apply Sternberg's *additive factors* method we need to have at least two *levels* of the discriminative variable: an *easy* discrimination (e.g., red versus green) and a more *difficult* discrimination (e.g., two hues of blue). Four randomized groups of subjects are needed for the four experimental conditions: (1) **c**-reaction with *easy* discrimination, (2) **c**-reaction with *difficult* discrimination, (3) **d**-reaction with *easy* discrimination, and (4) **d**-reaction with *difficult* discrimination. The mean RT obtained in each condition can be plotted as shown graphically in Figure 2.3. Panel A shows the absence of interaction between the **c** and **d** RT paradigms and the different difficulty levels of the discrimination. The increase in difficulty level causes the same increment in mean RT for both the **c**- and **d**-reactions. In contrast, Panel B shows a marked interaction effect. The **c**-reaction shows a large difference as a function of difficulty levels of the required discrimination, whereas the effect of difficulty on the **d**-reaction is relatively small. Subjecting the data in one of the panels (A or B) to ANOVA tells us whether each of the observed independent effects in the experiment (i.e., mean differences between factors, levels, and the interaction of factors × levels) attains a high enough level of statistical significance to be more likely a reliable phenomenon than a fluke of chance.

It was because of such interaction effects and their variable size that Donders and others obtained inconsistent results from the simple insertion and subtraction method. It was too much at the mercy of the interactions between RT and relatively small changes in the experimental conditions to yield consistent measures of the hypothesized mental processes.

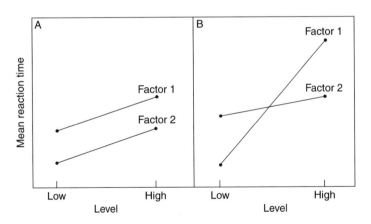

Figure 2.3: Graphical representation of the analysis of variance of the type applied by S. Sternberg (1969) to test Donders' hypothesis. Panel A shows a main effect for both factors (e.g., a difference between **c**- and **d**-reactions) and levels (e.g., low versus high discriminability of the RS) and there is no interaction between these main effects. Panel B shows the main effects of factors and levels and there is an interaction between them.

Whether an inserted effect is additive or nonadditive cannot be merely assumed. It is strictly an empirical question that must be answered by a proper analysis. Sternberg (1969), in fact, found some insertions to have almost perfectly additive main effects (like Panel A in Figure 2.3), whereas others showed significant interaction (Panel B).

The most usual interpretation of interaction effects attributes them to one or both of two conditions: (1) the addition of a complicating variable to the paradigm can change the nature of the task so that the times for the execution of processing stages in the simpler paradigm are no longer the same in the more complex paradigm, and (2) the cognitive demands of the inserted task may to some extent be processed simultaneously along with other stages of processing that are shared in common with the simpler paradigm, so the same segment of time supposedly representing a single-processing stage in one paradigm could accommodate more than a single process in some other paradigm. (Performing two or more distinct cognitive operations at the same time is known as *parallel processing*.). Donders' idea of a sequence of distinct stages — each representing a specific process that requires a certain amount of time — has long been virtually abandoned as a broadly valid generalization. It has figured the least in research on individual differences in RT, an application of mental chronometry that has focused more on the correlations of RT with various complex psychometric variables such as IQ, memory, and psychological effects of aging, health conditions, various drugs, and the like. Although the experimental study of variation in RT as a function of task complexity or variation in other stimulus conditions continues to be of interest, its modern theoretical treatment has not followed Donders (Pachella, 1974; Luce, 1986, pp. 473–483; Smith, 1980).

It should be noted that in using Sternberg's additive factors method based on ANOVA the data must consist of the raw RT measurements or their arithmetic mean. This a metrical advantage afforded by the true ratio scale and equal interval scale properties of RT. The additivity of experimental effects, in fact, cannot be rigorously or meaningfully tested unless the RT variable is measured in units of real time. Therefore any nonlinear transformation of the scale of RT should be eschewed. Also, the *median* RT should not be used for testing interactions in an ANOVA because medians are not necessarily additive, whereas the arithmetic means of values on a ratio scale or an interval scale are always additive.

## Elaborated and Formalized Chronometric Paradigms with a Timed Response

The most frequently used chronometric paradigms are described below under the names by which they are most commonly known. The descriptions are necessarily generic or abstract, because the paradigms are not standardized. For each paradigm many minor variations in apparatus and procedure are seen in the literature.

### Computerized Testing

Although most of the current paradigms are implemented by computers, this book does not deal at all with computerized testing *per se* — that is, when the computer merely substitutes for a conventional paper-and-pencil (PP) test. If a test's *item responses* are timed, however, it qualifies as a chronometric test. Item RT° can be a sensitive measure of item characteristics, particularly when, for the particular subject sample, most of the items are easy enough to be answered correctly within a relatively short time period. In data aggregated over a number of subjects, there is a strong positive relationship between response latency and item difficulty (measured as the percentage of the subject sample failing the item). Also, the response latencies on a set of test items that are so easy as to have zero errors in a given sample (e.g., college students) can predict the failure rates on the same set of items in another sample (e.g., elementary school children) in which there is a substantial failure rate. In other words, item response latencies can reveal differences in item difficulty levels even when every item has almost zero errors in the subject sample. The method has been used, for example, in measuring differences in difficulty in solving the simplest arithmetic problems, such as the time needed to respond either *True* or *False* to problems such as $2+2 = 3$, or $7-3 = 4$ (Jensen & Whang, 1993).

### *Binary RT Paradigm*

This is simply the presentation of RS to which P can give a binary response, such as Yes or No, True or False, Same or Different, etc. A number of other paradigms are built on this format. It has the advantage of keeping the response requirements of the RT task uniformly simple across wide variation in the difficulty of the discrimination task. The particular choice or discrimination RS is typically presented on a computer monitor, while P responds on a console with three keys or push buttons: a HK and two response keys for the binary responses. These binary response keys are clearly labeled (e.g., Yes–No), as shown on the response console in Figure 2.4. A wide variety of RS that allows binary discrimination has been used in this paradigm: purely visual or auditory sensory attributes, letters, words (e.g., synonyms–antonyms), numbers, abstract figures, faces, or simple arithmetic (e.g., the given answer is True or False). If E wishes to minimize the component of RT taken up by stimulus input time to obtain a purer measure of discrimination time, the

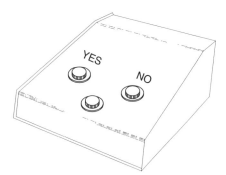

Figure 2.4: A binary response console ($6\frac{1}{2}'' \times 10''$) used for all types of binary responses (Yes–No, True–False, Same–Different, etc.). Magnetized labels can be quickly changed. The push buttons are $1''$ in diameter. The programmed response stimuli are presented on a computer monitor located directly above and behind the response console.

discriminative stimuli are presented successively rather than simultaneously and P responds to the second RS on each trial. If the successive RSs always appear at the same location on the presentation screen, it has the advantage of minimizing P's eye movements and diffusion of focus, which adds to the RT and thereby obscures to some degree the central-processing time for making the discrimination. P's RT is the interval between the onset of the discriminative RS and P's touching the correct button on the console.

### Sentence Verification Test

The sentence verification test (SVT) originated by Baddeley (1968) is a widely used form of the successive binary paradigm. A sentence appears on the screen describing a picture; then, after an interval that would allow the subject to read the sentence, the sentence disappears and a picture appears. P responds to the Yes or No key on the console as to whether the picture matches the verbal description. This test has also been called *semantic verification* because the descriptive sentence is often reduced to the bare essentials; for example, a simple preposition between two nouns. The best known is "Star over Plus" followed by a picture of a ★ directly above a ✚. On successive trials every possible permutation ($2^3 = 8$ in all) of the sentence is randomly presented (over, under, not over, not under) and the order of the words *star* and *plus* are randomly interchanged, as are the ★ and ✚.

A more complex version of the SVT is based on the first three letters of the alphabet, **ABC** (Jensen, Larson, & Paul, 1988). The verbal descriptors involve every possible difference in letter order, with a given letter described as first/last, or before/after a given letter, or between two other letters, and the negation of each of these conditions. The "picture" is always some permutation of the order of the letters **ABC**. The whole test consists of 84 trials comprising each of the 84 possible permutations of these conditions presented in a randomized order without replacement. (This form of the SVT has also been used as a nonspeeded PP test of verbal comprehension for elementary schoolchildren.)

# RT Tests Involving Retrieval of Information in Long-Term Memory

### Posner Letter-Matching Test

This test paradigm, originated by Posner, Boies, Eichelman, and Taylor (1969), is intended as a measure of the speed of retrieval of highly overlearned verbal codes in long-term memory (LTM). It actually consists of two distinct discrimination tests. The difference between their mean RTs is the point of interest. The binary decision in every case is *Same* or *Different*. One task, labeled *physical* (P), requires discrimination (Same/Different) between two letters of the alphabet based solely on their *physical* properties, e.g., the two members of the following letter pairs are all physically different: Aa, Bb, AB, and ab, whereas the letter pairs AA, BB, aa, and bb are all physically the same. The other task, labeled *name* (N), requires discrimination (Same/Different) between two letters based solely on the *names* of those letters, disregarding their physical properties, e.g., the two members of the letter pairs AA, BB. Aa and Bb are all the same, whereas AB, ab, Ab, and Ba are all different. Discriminating the physical differences does not require the retrieval of information from LTM; it could be performed just as easily using unfamiliar figures of comparable complexity. Discriminating the name differences, however, requires retrieval of the names of the letters from LTM. The difference in mean RT between the *name* and *physical* discriminations (i.e., N–P) is a measure of the retrieval time for letter names. It is negatively correlated with SAT scores in college students (Hunt, Lunneborg, & Lewis, 1975).

### Synonyms–Antonyms Test

This variation of the Posner paradigm is based on easily recognized words. It was introduced as a more difficult and discriminating RT task for college students (Vernon, 1983). The *physical* task requires a Same/Different discrimination between word pairs that are physically the same or different (e.g., *Dog–Dog* versus *Dog–Log*). The *name* task requires a Same/Different discrimination based on the semantic meaning of the paired words, which are either *synonyms* or *antonyms* (e.g., *Bad–Evil* versus *Bad–Good*). As the aim of this test was not to test vocabulary but to measure the speed of retrieving word meanings from LTM, the words used as synonyms and antonyms were selected from the Lorge–Thorndike word list as familiar, high-frequency words. Under nonspeeded conditions, college undergraduates would show virtually error-free performance.

### Verbal Classification or Category Matching

This RT test also requires that the subject access information in LTM. P responds *Yes* or *No* according to whether a presented word (or picture of a familiar object) can be classified into some specified category. There are a number of procedural variations of this task. A typical application, for example, used five categories of words (animal, clothing, fruit, furniture, and sport). P was presented with pairs of words and was required to make a binary response (*Yes* or *No*) as to whether the second word in each pair was a member or a nonmember of the same category as the first word (Miller & Vernon, 1992). The test is especially suited to the developmental study of concept formation and categorical thinking, which are related to children's mental age and IQ.

### The Sternberg Short-Term Memory Scanning Paradigm

This famous paradigm was first presented by Saul Sternberg (1966, 1969) and attracted considerable attention because of the apparent simplicity of the findings, which proved to generalize across many variations of the procedure. (*Note*: There are two well-known Sternbergs in this field, Saul and Robert J., so it is often useful to denote them by their initials.)

The essential paradigm consists of presenting P with a set of 1–7 items (digits, letters, or symbols) selected from a larger pool of items (e.g., digits from 0 to 9). The items presented on a given trial are called the *positive set*. After a brief interval (2 s), the positive set is replaced by a single *probe* digit, which was either a member or not a member of the positive set, and P responds accordingly by pressing one of the two response keys labeled *Yes* and *No*. The *set size* (i.e., number of items) is varied across trials. RTs increase linearly as a function of set size. Sternberg presented the items *sequentially* at a rate of 1.2 s per digit; 2 s after the last digit, a warning signal appeared, followed by the probe digit. Other studies have used *simultaneous* presentation of the positive set (e.g., digits in a horizontal row) for 2–3 s. Each additional item in the set requires the same increment of processing time, regardless of the probe digit's position in the set; that is, the difference between the RTs for set sizes 1 and 2 is the same as the difference between RTs for set sizes 5 and 6, regardless of the position of the probe digit in the positive set. This finding indicates that the memory-scanning search is *exhaustive*; that is, P scans the whole positive set to the last item in the series, even when the probe item is the first item in the series. The same phenomenon is seen whether the items are digits, letters, or symbols. Figure 2.5 shows the result of Sternberg's (1966) classic experiment based on only eight subjects. In other studies based on much larger samples (e.g., Jensen, 1987a) it is clear that the mean RT for negative responses is consistently longer than for the positive responses, as shown in Figure 2.6.

### The Neisser Visual Scanning Paradigm

Attributable to the cognitive psychologist Ulrich Neisser (1967), the procedure in this paradigm is the exact opposite of Sternberg's memory scanning (MS) paradigm. Visual scanning (VS) makes no demand on MS but measures the speed of scanning a set of digits or letters to determine whether or not the set contains a previously displayed probe. First, a single probe digit is presented for 2 s, followed after 2 s by a series of 1–7 digits presented simultaneously. P visually scans the series for the presence or absence of the probe digit and responds by pressing either the *Yes* key or the *No* key. The results of VS are highly similar to those for MS, although the RTs for VS are slightly longer than for MS, as shown in Figure 2.7. The similarity between MS and VS seen in Figure 2.7, as well as the similarity of their correlations with various other chronometric and psychometric variables suggests a psychophysical isomorphism between the mental and VS processes. Information in short-term memory (STM) is apparently processed in much the same manner as information presented in the visual field.

There are more recent visual search tasks that have interesting correlations with psychological variables. The *k*-test is described as "a computerized continuous performance task in which subjects decide if the letter *k* is present among 10 distractor characters" (Dalteg et al., 1997). There are two conditions based on the nature of the distractors: (1) filled squares and (2) other letters. The letter *k* appears randomly among the 10 distractors on 50 percent of the trials. On each trial the distractors are scattered on the computer monitor in a different random constellation. P has to respond on *Yes* or *No* keys whether the display contains the letter *k*. Performance under each condition is scored for both mean RT and accuracy.

### Continuous Recognition Memory

Introduced as an RT paradigm by Okada (1971), P is presented with a list of familiar words as the memory items, displayed sequentially at a constant rate. P is instructed to recognize whether the particular word being presented has or has not been presented earlier in the sequence (often referred to as *old* or *new* words, respectively). As each word appears, P presses one of the two keys (*Yes* or *No*) identifying the word as having been seen previously (*old*) or not been seen before (*new*). RT to the *old* items is plotted as a function of the degree of *lag* (i.e., number of intervening items) between the earlier and later presentations of a

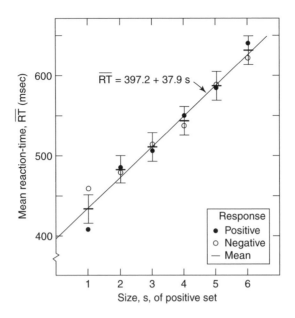

Figure 2.5: Mean RT as a function of set size of the positive set. From Sternberg (1966). (Reprinted with permission of *Science*.)

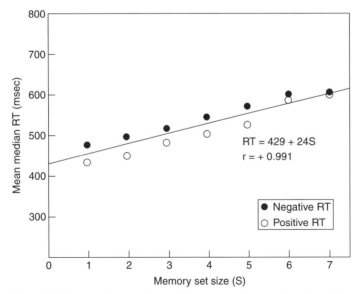

Figure 2.6: Mean RT for positive and negative responses as a function of set size in the memory scan test. The regression line is based on the means of the positive and negative responses on set sizes 1-7. From Jensen (1987a). (Reprinted with permission of Elsevier.)

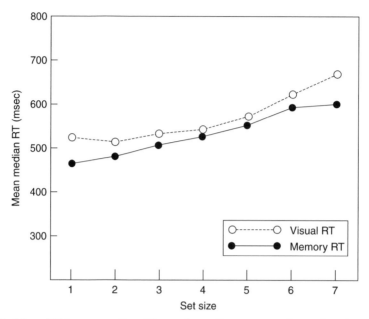

Figure 2.7:  Mean RT (average of positive and negative responses) as a function of set size on visual scan and memory scan. From Jensen (1987a). (Reprinted with permission of Elsevier.)

particular word. The number of words in the list, the number of repetitions of a given word, and the lag intervals are varied in accord with E's purpose. Compared with most other RT paradigms, typical RTs in this test are relatively slow (600–800 ms in the simple form used by Okada). Recent research has shown it to be a highly sensitive measure both of STM and of resistance to the interference of the intervening lag items.

**Coincidence timing (CT)**   CT involves more than simple RT as it requires something like anticipation, estimation, and expectation as well as quickness of response at the point where a horizontally moving dot coincides with a thin vertical line in the center of the computer monitor. In the paradigm first described by Smith and McPhee (1987), the dot leaves the right or left margins of the screen at random intervals, and there are three conditions, each given on 30 trials: (1) a straight horizontal path of the dot's movement at a speed of 0.10 m/s, (2) the same as (1), but with a movement speed of 0.15 m/s, and (3) a randomly jagged path of the dot's movement at 0.10 m/s. P responds to the dot/line coincidence by pressing the space bar on a computer keyboard. Performance on each condition is measured as the mean absolute distance between the line (true position) and the position of the dot when P responds (response position). (Note that this error measured as distance can be converted into a time measure of error, given the speed of the moving dot.) Another performance measure is the SD of the differences. These measures are found to be moderately correlated with conventional psychometric measures of intelligence (Smith & McPhee, 1987; Larson, 1989).

**Analogical decomposition paradigm**   This quite specialized method, originated by R.J. Sternberg (1977), has been used to analyze the various "cognitive components" in relatively complex mental tasks. A *cognitive component* is a hypothetical construct defined as an elementary information process that operates upon internal representations of objects or symbols. One or more of a limited number of such components is presumed to be involved in virtually every kind of cognitive task. The precise amount of time taken by each of the specific components hypothesized to be involved in the performance of a given task is measured as a binary CRT. The method is applicable, for example, to analogy problems typical of those often used in IQ tests. It aims to measure individual differences in various elemental cognitive components.

According to Sternberg's theory, the total solution time for an analogy problem consists of the separate times required for components such as encoding the various terms in the analogy problem, abstractly represented as A:B::C:D. The components (in italics) involved in such an analogy are *encoding* each of the elements in the analogy, *inferring* the relationship between A and B, *mapping* the relationship between A and C, and *generating* (or recalling from LTM) a word that has a comparable relationship between C and D. The D term in the actual analogies presented is always selected by P from a binary (True or False) choice in which only one of the two alternatives fits the analogy, and P's CRT is measured. The decomposition of the component times is accomplished by presenting the elements of the analogy in stages, allowing P all the time needed to process (i.e., encode) each successive element before presenting the whole analogy. For example, P looks at A for as long as necessary to encode it, then presses a key that brings up B::C:D and makes a choice response as quickly as possible consistent with accuracy of response. Hence only the CRT to process the B::C:D part of the analogy is measured. Over a number of analogy items, the mean CRT for processing the full analogy (A:B::C:D) minus the mean CRT for processing a part of it (B::C:D) is a measure of the encoding time for A. Then A:B is presented for as long as P needs, followed by C:D, and subtracting the CRT for C:D subtracted from the total time for the full analogy gives a measure of the time taken to encode A and B and to map the relationship between A and B. Thus by measuring the CRT for each of the successive remaining elements of the analogy and with the appropriate subtractions of the CRTs from the total CRT and from one another, it is thought possible to decompose the total time for performing the analogy into its constituent segments. Whether individual differences in a hypothesized component measured as a time segment derived by the subtraction method is invariant across different tasks is problematic. Sternberg's method is also liable to the same problems pointed out by the critics of Donders' subtraction method.

Unfortunately, the early promise of this method of componential analysis was unfulfilled, mainly because of inconsistencies of empirical results and certain critical failures of replication. Individual differences in specific components did not generalize much across similar tasks in which they were hypothesized to operate. The common factor across various components was the only source of the individual differences variance that consistently correlated with IQ, while the different cognitive components independent of their common factor were largely task-specific. This condition makes for poor convergent and discriminant validity; that is, the measured components show little of the expected generality across tasks that supposedly involve the same hypothesized components, and the measured components correlate about as much with tasks in which they are hypothetically absent (Brody, 1992, pp. 103–124; Lohman, 2000, pp. 313–314). It should be noted, however, that these deficiencies are entirely on the theoretical side, involving models of information processing in terms of various hypothetical cognitive components. The method *per se* should not be dismissed on these grounds alone. The nontheoretical or practical validity of the chronometric data yielded by this method must depend on its empirical correlations with other variables of interest.

### Spatial Rotation Paradigm

This test, originally devised by Roger N. Shepard and colleagues, requires the mental rotation of still images projected on a screen. It is a highly sensitive measure of facility in the mental visualization and rotation of objects. There are two main versions. In one version, the images are projected as three-dimensional (3-D) objects (Shepard & Metzler, 1971). A simpler version projects 2-D abstract figures or alphanumeric characters (Cooper & Shepard, 1973; Cooper, 1976).

The essential paradigm is most easily illustrated by the 2-D version. It is a DRT test. P is shown a pair of figures (or letters) that appear simultaneously on the screen; the second item in each pair is the same as the first item but has been spatially rotated (in the same 2-D plane) by a certain number of degrees ($0° - 180°$) from the position of the first figure. P responds on keys labeled *same* or *different*, depending on whether the second figure is the *same* as the first in the sense that the second figure could be made perfectly congruent with the first simply by rotating the second figure in the same plane. P responds *different* in a random half of the trials where the second figure is a mirror image of the first and therefore could not be made congruent with it by any degree of rotation in the same plane (Figure 2.8).

The P's DRT is the average time needed to make this discrimination between congruent (*same*) and incongruent (*different*).

The theoretical importance of the Shepard et al. experiments is that the DRT is found to be a linear function of the number of degrees that the second figure must be rotated to make it congruent with the first figure. This fact indicates a psychophysical isomorphism between the external figure and its internal representation as it is mentally rotated through the distance (*n* degrees) that the actual rotation the physical figure would have to undergo to be congruent with the target figure. This linear relationship between RT and degrees of rotation holds for nearly all individuals as well as for group data (Cooper, 1976). Moreover, there are reliable individual differences in the slope of this linear function. This

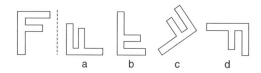

Figure 2.8: Rotated figures (e.g., the target letter **F**), where figures *a* and *d* are rotated in the same plane; *b* and *c* are turned over in 3D space and are also rotated. RT can detect the different speeds with which these two types of rotations can be recognized.

phenonenon has proved of interest to differential psychologists because of its relationship to individual differences in spatial visualization ability as measured by a variety of PP tests and provides a more finely graded analytic measure for studies of the nature of spatial visualization abilities (Pellegrino & Kail, 1982).

### The Hick Paradigm

One of the most widely used paradigms in research on information processing, this paradigm is named after a phenomenon now commonly referred to as *Hick's law*. In a classic CRT experiment performed by W. E. Hick (1952), a series of nine CRT tests is presented in which there are anywhere from 1 to 10 different degrees of choice (*n*); that is, the number (*n*) of RS differs on each task in which there is a corresponding number (*n*) of response alternatives. The different RS are presented at random. Hick showed that the CRT increases by a constant amount according to the amount of uncertainty as to which one of the RS would appear next. Uncertainty is measured in *bits* (for binary digits). In information theory, one *bit* is defined as the amount of information that reduces uncertainty by one-half. In the Hick experiment it is the binary logarithm of the number of stimulus–response alternatives (*n*); that is, CRT increases by constant increments as a function of $\log_2 n$. Then, $RT = a + b \log_2 n$, where *a* is the intercept and *b* the slope of the linear function. This formulation is known as *Hick's law*. The intercept (*a*) is theoretically measured by SRT, which with only one stimulus and one response, assumes zero uncertainty. However, there is always some uncertainty as to precisely when the RS will appear. To take account of this, Hick's law is given as $RT = a + b \log_2 (n + 1)$, which sometimes has a slightly better fit to the data. Figure 2.9 illustrates Hick's law fitted to a set of CRT data obtained by Merkel (1885). The slope (*b*) of the Hick function is said to be a measure of the "rate of gain of information."

Roth (1964), a German psychologist, had the idea that the "rate of gain of information" (i.e., the slope parameter, *b*, of the Hick function) should be related to intelligence. He published a study that substantiated this hypothesis, thereby bringing Hick's law to the attention of differential psychologists (e.g., Eysenck, 1967). But it was not further investigated from this standpoint until some years later when it was studied empirically with a modified version of Hick's apparatus, which later came to be dubbed the "Jensen box" (Jensen & Munro, 1979). There followed many other studies by Jensen and colleagues and by other investigators using a similar apparatus. This apparatus has been described in detail elsewhere, along with an exposition of all the main findings of 33 studies of individual differences in the Hick paradigm (Jensen, 1987b). The original apparatus is shown in Figure 2.10, with all eight of the light/button alternatives exposed. The set size of the number of stimulus/response alternatives can be varied by placing over the console a template that covers any given number of the light/button alternatives. The set sizes typically exposed in RT experiments are 1, 2, 4, 8 corresponding to 0, 1, 2, 3 bits of information. In a later version of the console, instead of having lights and push buttons, there were only green push buttons, 2 cm in diameter, any one of which lights up as the RS and goes "off" when touched. Having only the underlighted push buttons insures the maximum stimulus–response compatibility, which has proved to be an especially important feature in the testing of young children and the mentally handicapped.

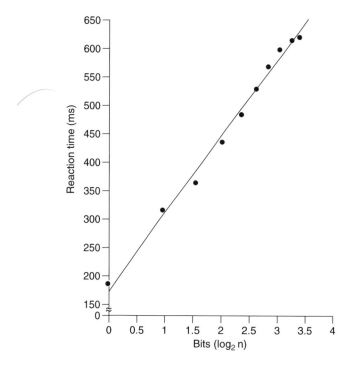

Figure 2.9: Mean CRTs to stimulus arrays conveying various amounts of information scaled in bits, where *n* is the number of choice alternatives. Data from Merkel (1885) as reported in Woodworth and Schlosberg (1954, p. 33). (Adapted with permission from Holt, Rinehart, & Winston.)

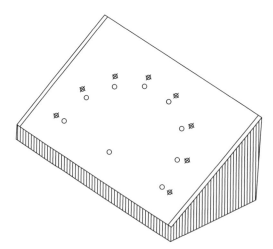

Figure 2.10: Response console of the Jensen box. Push buttons are indicated by circles, faceted green lights by crossed circles. The home button is in the lower center, 6″ from each response button.

The testing procedure begins with P's index finger on the black HK on the lower center of the console. One or two seconds later a PS sounds ("beep") for 1 s, followed by a continuously random interval from 1 to 4 s, whereupon the RS lights up. As quickly and accurately as possible, P releases the HK and touches the lighted RS button to turn it off. Two time intervals are automatically measured and recorded: RT and MT. These measures are relatively independent, correlating with each other only about .30. Whereas RT increases with set size according to Hick's law, MT remains approximately constant across all set sizes, as shown in Figure 2.11. Apparently the processing of the RS takes place entirely before P leaves the HK. To make it difficult or impossible for P to leave the HK before completely processing the precise location of the RS, some researchers have all of the lights go "on" the instant P releases the HK, in order to lessen the possibility for P to leave the HK prematurely and continue processing the location of the RS while in transit (Smith & Carew, 1987). This "masking" of the RS at the instant P responds is intended to discourage response strategies that interfere with conformity to Hick's law. It is important in RT experiments to ensure to the fullest extent possible that all Ps are actually performing the same task in the same way, rather than adopting different strategies. Although this procedural variation actually makes little difference for the vast majority of Ps, I believe it is nevertheless an improvement and I strongly recommend it for all future applications of the Hick paradigm.

One variation in procedure that I believe should be avoided dispenses with the HK and has the P place the four fingers of each hand on a row of eight push buttons (or a standard computer keyboard). P responds to each of the RS lights by pushing a corresponding button. This procedure confounds the RT with manual dexterity, differences in handedness, and in the strength and agility of the different fingers (which, as every pianist knows, is considerable).

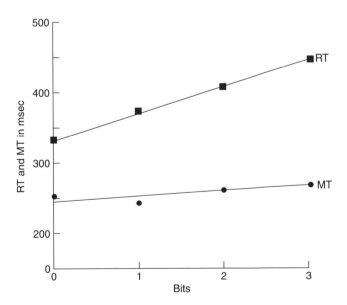

Figure 2.11: RT and MT as a function of the "information load" scaled in bits (the binary logarithm of the number of response alternatives). Data from Jensen (1987b).

Another procedural variation has been devised to minimize *retinal displacement* effects, which have been thought to affect the RTs in the Jensen box procedure described above. There is the possibility that P's eye movements across the response console to focus on the RS when it goes "on" may add different increments to the RT depending on the location of the particular RS. To minimize this irrelevant source of RT variation, Neubauer (1991) devised an apparatus (see Figure 2.12) with four light/button alternatives, in which the lights (green LEDs) are closely adjacent to the response buttons and are arranged around the HK in a 3 cm semicircle — an arrangement making for a high degree of S–R compatibility and very little VS of the console. All stimulus information is masked the instant P releases the HK.

Another arrangement of the same apparatus described above, devised by Neubauer (1991), has a lower degree of S–R compatibility, but completely eliminates VS, or retinal displacement. The four LEDs adjacent to the response keys consistently display the numbers 1–4. A larger LED, which displays the RS, is located above the LEDs and response keys. The large LED displays a number (from 1 to 4) to which P responds by pressing the response button adjacent to the LED displaying the corresponding number. It had been claimed by some critics of the Jensen box procedure that the correlations between IQ and the Hick variables (RT and RTSD) resulted from peripheral "strategy" artifacts created by individual differences in retinal displacement, S–R compatibility, and the like, rather than representing a correlation between IQ and central processing speed *per se*. Neubauer's (1991) modification of the Hick task intended to minimize or eliminate these potential artifacts revealed that, if anything, such effects only attenuated the correlations between IQ and the Hick RT variables.

### Odd-Man-Out Paradigm

This paradigm transforms the Hick paradigm from being a form of CRT to being a form of DRT. The Odd-Man-Out (OMO), as it was named by its originators (Frearson & Eysenck, 1986), was devised specifically to complicate the Hick paradigm, with the expectation of

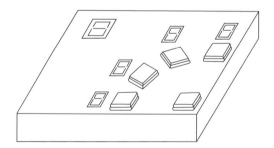

Figure 2.12: Neubauer's RT/MT apparatus for CRT. The home button is in the lower right-hand corner, equidistant from each of the four response buttons. Adjacent to each response button is a green LED digital display. In the upper left corner is a larger red digital display LED, which presents the response stimulus (RS). (Reprinted from Neubauer, 1991, with permission of Elsevier.)

increasing its correlation with IQ. The same apparatus used in the Hick setup (Figure 2.10), with all eight push buttons exposed, is used in OMO. The essential difference is that, on each trial, three of the eight lights go "on" simultaneously. Two of the lights are always closer together than the other, which is the "OMO." P's task is to touch the OMO button as quickly as possible, which instantly turns off all three of the lighted buttons. With eight lights, there are altogether 44 possible OMO combinations, one of which is shown in Figure 2.13, and all 44 of these combinations are presented in a random order without replacement. In every other way the procedure is exactly the same as in the Hick paradigm with eight buttons. Having to discriminate the distances between the three lighted buttons increases the mean RT about 200 ms over the average RT for the eight-button Hick task, and the correlation of OMO's mean RT with IQ is nearly double that obtained with the eight-button Hick task.

### Dual Task Paradigms

Dual tasks are another way of complicating an RT paradigm. It taxes P's working memory capacity to some extent (Pashler, 1993). Working memory (WM) is a hypothetical

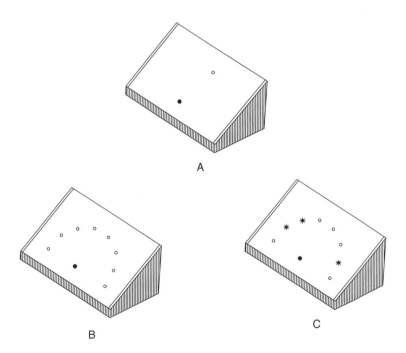

Figure 2.13: The response console for (A) SRT, (B) CRT, (C) DRT (odd-man-out). The black dot in the lower center is the home key (HK). Six inches from the HK are green, under-lighted push buttons. In the SRT and CRT conditions (A and B) only one button lights up at random on each trial. On the DRT (shown in C), three buttons light up simultaneously on each trial. The distances between the three buttons are unequal; the button remotest from the others is the odd-man-out.

construct used to account for one's capacity to retain information in STM for a brief period while having to deal with an additional attention-demanding stimulus input. For example, a dual task combines two distinct RT paradigms, say, (1) *memory scan* for a series of numbers, and (2) *synonym–antonym* (**S–A**) discrimination. The sequence of events (with the displayed stimuli in bold face) and the automatically controlled time intervals are as follows:

| Events | Times |
|---|---|
| Preparatory signal ("beep") | 1 s |
| Blank screen | 1 s |
| **84135** | 3 s |
| **LARGE–SMALL** | |
| P presses button **A** | $RT_1$ and $MT_1$ |
| Blank screen | 1 s |
| **3** (probe digit appears) | |
| P presses button **YES** | $RT_2$ and $MT_2$ |

The effect of the dual task (i.e., having to retain the number set while making the S–A discrimination) considerably increases both $RT_1$ and $RT_2$ Each of the dual task RTs is also more highly correlated with psychometric *g* than is either one of the tasks when presented separately, presumably because of the additional strain on WM capacity in the dual task condition.

Dual tasks can also be presented simultaneously such as presenting at the same time different auditory and visual stimuli requiring some form of discrimination. For example, P must respond whether two successive tones are the same or different in pitch while discriminating whether two visually presented words are the synonyms or antonyms. The responses, of course, have to be made separately in time. Dividing P's attention by such procedures that involve "parallel processing" of information markedly increases RT and also generally increases its correlations with untimed psychometric measures of cognitive ability. One problem with simultaneous dual tasks, as compared with single tasks, is that dual tasks are more liable to elicit different performance strategies, which may have different effects on RT, thereby obscuring the causal basis of its correlation with *g* or other variables.

## Chronometric Tasks without Timed Response

As already noted, RT has two major aspects, the sensory and the motor, involving the afferent and efferent neural pathways and the complex network of connections between them in the brain. The RS sets off a physioloigical activity in the *sensory* and *motor* pathways and their neural connections that produce the final RT. If one side of this *sensory–motor* system is less closely related to the information-processing functions of the CNS than the other, it acts as an attenuating factor in the correlation of RT with cognitive ability, and its contribution to a timed measured of cognitive performance might better be minimized or avoided altogether. This is attempted statistically in RT studies by regressing SRT out of more complex forms of RT, such as CRT and DRT (Jensen & Reed, 1990). It would perhaps be more advantageous to have a measure of information-processing time that is dependent only on the afferent system, completely excluding the efferent component from the time measurement. The following paradigms meet this condition. Here, following the presentation of the RS, the measured times are all on the intake side; motor RT is not of interest, and is measured only incidentally, if at all.

### Inspection Time

The inspection time (IT) paradigm, attributable to the Australian cognitive psychologist Douglas Vickers, has become one of the most remarkable and well investigated chronometric paradigms (Vickers, Nettelbeck, & Willson, 1979). In the typical form of the IT test the discriminative stimulus consists of two parallel vertical lines ("legs") connected by a horizontal line at the top. One leg is twice the length of the other. After a visual fixation point appears on the screen, the pair of legs appears, with the longer leg randomly on either the right or left. This figure is quickly followed by a *masking stimulus* that completely covers the test stimulus to prevent further processing of it. The test stimulus (a) and the mask (b) are shown in Figure 2.14. P holds a thumb button in each hand and responds by pressing either the left or right button on each trial to indicate whether the longer leg appeared on the right or left side. E's instructions to P carefully emphasize that this is not a RT test but a measure of perceptual speed, and that P should take as much time as needed to decide which of the two legs is the longer. P is required to make a "best guess" when uncertain. The testing procedure is run automatically by an adaptive computer program, which systematically varies the time interval between the onset of the test stimulus and the mask, changing the duration of the test stimulus in small steps in response to P's performance. P's correct responses are followed by a shortening of the interval on the succeeding trial, incorrect reponses by a lengthening. As the chance guessing score is 50 percent correct responses, the computer program automatically zeroes in on the presentation duration at which P attains the equivalence of 97.5 percent correct responses, a performance criterion defined in the IT literature as $\lambda$ (lambda), i.e., the duration of the test stimulus needed to attain a 97.5 percent probability of a correct discrimination. Individual differences in $\lambda$ range widely, depending mainly on the age and intelligence of the subjects (given equal practice). The values of $\lambda$ are distributed

Figure 2.14: The best known form of the inspection time (IT) paradigm. The figure labeled **a** is presented for a brief period (e.g., 70 ms) and is immediately followed by **b**, which covers and 'masks' the retinal after-image of figure **a**. The interval between the onsets of **a** and **b** needed for a person to report with 97.5 percent accuracy whether the longer "leg" of figure **a** was on the left or the right side is termed that person's IT. This measure of IT is conventionally symbolized as $\lambda$.

around a mean of about 100 ms. One of the most remarkable facts about IT is that $\lambda$ has a consistently higher correlation (about $-.50$) with IQ than any measure of a single parameter derived from any other chronometric paradigm (Kranzler & Jensen, 1989).

Researches on IT are especially particular about the apparatus used for the IT task and frown upon presentation of the test stimuli by computer monitors because of the relatively slow "refreshment time" of typical monitor screens, which is about 10–15 ms, and the lack of a perfectly square on/off wave for the stimuli presented. These features of ordinary computer monitors can create artifacts in the precise measurement of $\lambda$. Therefore, the use of LEDs is the preferred method for presenting the IT test. Figure 2.15 shows the LED apparatus used in one of the leading laboratories for IT research, at the University of Edinburgh (Egan & Deary, 1992).

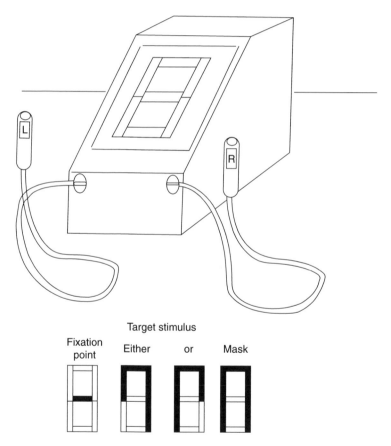

Figure 2.15: An apparatus for presenting the IT paradigm shown in Figure 2.14. The target stimulus and the mask are presented by means of LEDs. P responds by pressing the left or right thumb button (held in the L and R hands) to indicate whether the longer of the two IT "legs" were presented on the right or left of the display. (Reprinted from Egan & Deary, 1992, with permission of Elsevier.)

Some small percentage of Ps taking the IT test report experiencing an after-image, or an apparent motion illusion of a rapid downward extension of the short leg at the instant the test stimulus is covered by the mask. They may develop a strategy of using these clues in performing the IT test. (In general, strategy use has not been found to affect $\lambda$ much or its correlation with other psychometric variables independent of other P variables.)

Variations of the IT procedure have been devised to eliminate these visual artifacts. In one such method, the test stimulus consists of a perfect 3 cm² with one of its four sides missing (the missing side is randomized over trials); the mask consists of a complete square that covers the test stimulus. P responds by pressing one of the four buttons to indicate which side of the square was missing (Zhang, 1991). The IT criterion was an 85 percent accuracy rate, which, in 40 university students corresponded to a mean stimulus duration of 54 ms with *SD* of 16 ms and a range from 18 to 90 ms. The IT measures correlated −.71 with IQ (Raven Standard Progressive Matrices).

Another IT task found to eliminate the visual artifacts and strategies that affect the two-line IT for some Ps is entirely computer administered with the IT stimuli projected on a Super VGA monitor. The test consists of presenting pairs of letters which are either the same or different, followed by a mask of five different letters that fill the space previously occupied by the stimulus pair, to which P indicated "same" or "different" on a keyboard (Stokes & Bors, 2001).

The previous study also illustrates another common method of estimating IT that has found use in a number of computer applications because of its simplicity. It does not require an interactive program that continually controls the stimulus duration intervals in accord with P's trial-to-trial performance. Hence the program does not reactively "zero-in" on P's IT. As a result, it forsakes some degree of efficiency and precision. The exposure time of the discriminative stimulus is predetermined by E in the computer program. A set number of variable stimulus exposure times, ranging between the lower and upper limits of perceptual speed in the tested population is presented with the same number of trials for each exposure time. P's responses are scored either as the total number of correct discriminative responses made over all of the exposure times or as the length of the specific exposure time on which P makes some specified number (or rate) of correct responses, e.g., 19 out of 20 trials. P's IT is estimated as the midpoint of the exposure time interval in which the specified criterion is attained. Mathematically more complex methods for estimating IT are presented elsewhere (Deary & Stough, 1996; Barrett, Petrides, & Eysenck, 1998).

**Auditory IT**  This is an analog of visual IT, except that all the stimuli are auditory and the discriminative stimuli must therefore be presented sequentially instead of simultaneously (Raz, Willerman, Ingmundson, & Hanlomnf, 1993). First, it is established that P could easily discriminate between a high (H) tone and a low (L) tone. Then, on each trial, a pair of H and L tones (each with a duration of 20 ms) is presented sequentially in a random order, followed by a masking stimulus (e.g., a tone pitched halfway between the H and L tones) of 500 ms duration. (The H and L pitches and the mask pitch are constant throughout all trials.) The ISI between the H and L tones is varied across trials by increments from 0 to 1000 ms. On each trial, P indicates whether the first tone of the pair is high or low. In a college population, a 95 percent correct response rate occurs at an ISI between 100 and 200 ms. Like visual IT, auditory IT has a remarkably substantial correlation with IQ and SAT scores. It is not correlated with musical training.

### Visual Change Detection

This variation of the IT paradigm was devised to broaden the generality of measuring the IT phenomenon beyond the specific visual-processing task originally used to measure IT, and to determine if a somewhat different kind of IT task is also comparably correlated with IQ and other psychometric abilities (Deary, McRimmon, & Bradshaw, 1997). The task is most easily explained by Figure 2.16. A PS (visual fixation cue) appears for 100 ms in the center of the screen of a color, touch-sensitive computer monitor. There follows a complex pattern of 49 squares (49 of the squares in a 10 × 10 grid), presented across trials with variable durations of six evenly spaced IT durations, from 14.3 to 85.7 ms. This is immediately followed by a second pattern which is the same as the first except that it now contains 50 squares. P responds, taking as much time as needed, by touching the added square on the touch-sensitive screen. If the exposure time of the target stimulus was too short, P cannot identify the added square, but if the exposure time is long enough, the added square tends to "pop out" at P. A variation on this procedure, called *visual movement detection* (VMD) is to have one of the squares move right or left by a distance equal to its width after a variable time. The VMD task is much easier than the visual change detection (VCD). The original IT paradigm and VCD and VMD all share a large common factor, with factor loadings of IT = 0.57, VCD = 0.78, VMD = 0.81. The common factor among these tests was found to be correlated .46 with the common factor extracted from four tests of verbal and numerical abilities.

### Frequency Accrual Speed Test

Originated by Douglas Vickers (1995), the frequency accrual speed test (FAST) derived from the same theory that led to his invention of the IT paradigm, viz., that individuals differ in their speed and efficiency of processing the stream of information delivered by the

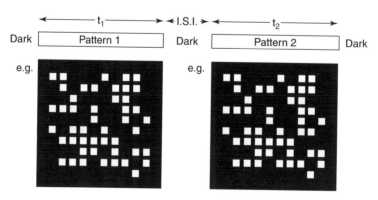

Figure 2.16: Stimulus displays used in appearance and disappearance detection studies. (From Deary et al., 1997, with permission of Elsevier.)

senses, and the faster the rate of information flow, the greater the probability that some of the information does not make it through the uptake system and is lost for further processing, as is needed for its short-term retention, for example. It is problematic to classify this paradigm strictly as a form of IT, however, because it was later concluded by Vickers himself and by others (Deary, 2000a, pp. 203–206), that FAST performance reflects supra-span memory as much or more than the speed of processing. ECTs differ on the simplicity–complexity continuum of the number of different information processes they tap, and in this respect FAST appears more complex than the typical IT paradigms previously described.

The FAST essentially comprises the following succession of events: on each of a given number of trials two left–right adjacent LEDs flash sequentially in a random order for a fixed number of flashes. At the end of the sequence, P responds by pressing buttons labeled Left or Right to indicate which LED had flashed more frequently. The relative frequencies of the R and L flashes were in the ratio of 17:13, and the more frequent LED was randomized with respect to its R and L positions across trials. The number of flashes in a trial far exceeds anyone's capacity to count and remember the exact number of flashes on the left or the right, but as the number of flashes increases, P accrues an impression that there were more flashes on one side than on the other. E can systematically vary the flash rate. At higher rates more information is lost, reducing P's probability of a correct judgment at the conclusion of each trial. Several rather minor procedural variations of FAST are described in the references previously cited. Individual differences in accuracy on FAST are correlated with IT and also have shown correlations with psychometric intelligence ranging from .50 to .74 (Caryl, Deary, Jensen, Neubauer, & Vickers, 1999).

### Eye Movements

Professor Niels Galley, a neuropsychologist in the Psychological Institute of the University of Cologne, Germany, has used certain features of eye movements as a chronometric method for the study of IQ and *g* (Galley & Galley, 1999). In 1998, I visited Prof. Galley's laboratory in Cologne to learn first-hand about his innovative research. Except for the fact that it calls for quite specialized instrumentation and computer programs seldom found in psychological laboratories, the method itself is remarkably simple and efficient.

Galley's procedure essentially measures *saccades* and *fixations*. Saccades are the very short, quick, and irregular jerky movements of the eye when shifting its gaze from one fixation point to another. Saccades occur spontaneously at the rate of about four per second but can be driven by a rapidly moving visual stimulus that the subject is required to follow. Hence a great quantity of the essential data can be obtained in a short time. The *fixation duration* (FD) is the interval between the termination of a saccade and the beginning of the next saccade. FDs typically range between 100 and 300 ms and rarely exceed 500 ms. The FDs can be measured and recorded to 1 ms accuracy by an electrooculogram, from which the data are fed into a computer for quantitative analyses, which includes the mean, median, and *SD* of P's FD measured under different stimulus conditions. These measures are obtained for FD with eyes closed, FD during a relaxed interview, FD while carefully examining a life-like picture, FD while solving problems on the Raven Matrices, and FDs while performing two visual tasks, one requiring the eyes to follow a left–right jumping light at various "jumping" speeds, the other requiring the eyes to follow a spot of light wandering in a wavy motion horizontally across the screen of a computer monitor. Significant negative correlations of −.25 to −.30 with IQ were found only on the visual tasks that brought forth a quickening of the FD, thereby increasing the rate of fixations while inspecting the given stimulus. The tasks that drove the FD rate the most and thus were the most highly correlated with IQ were the Raven test and the jumping light. (The Raven was not used as an IQ test, but only as a visual stimulus for measuring FD. The test used to measure IQ in this study was the German version of the Wechsler Adult Intelligence Scale.) It is not so surprising that these eye movement phenomena are related to IQ, because the saccadic system is amazingly complex, being under the control of many brain loci, including the brain stem, cerebellum, and cerebral cortex (Glimcher, 1999). Therefore, it may be that the saccadic variables measured in Galley's experiment are highly sensitive to the overall intactness and "fine-tuning" of the brain's higher neural functions, which are more globally reflected by psychometric *g*.

### The Stroop Color–Word Test

This test has fascinated psychologists in various fields ever since its invention for use in the doctoral dissertation research by J. Ridley Stroop (1935). It has since accrued an extensive research literature (Jensen & Rohwer, 1966). The Stroop test has been generally regarded as a measure of an individual's resistance to interference due to response competition, as when the execution of an $S_1$–$R_2$ connection is blocked or delayed because of interference from a prior learned connection, $S_1$–$R_1$. For this reason, it has been of interest in the psycholinguistic study of bilingualism and various types of interference in speed production. Also, experimental cognitive psychologists have used chronometric techniques to test hypothetical models of brain action intended to explain the various features derived from the Stroop effect (Posner, 1978, pp. 91–96).

The Stroop test consists of three distinct tasks. On each one, P's performance is timed in seconds with a stopwatch or chronometer.

A. *Color naming*: P is shown a chart on which there are several lists of small rectangles, each one solidly colored either red, green, yellow, or blue, in a pseudo-random order. P is told to name the colors of all the rectangles in consecutive order as quickly as possible.
B. *Word reading*: P is shown a similar chart on which are lists of the same number of color *names* (red, green, yellow, or blue), all of them each printed in black ink, and P is told to read aloud the printed words as rapidly as possible.
C. *Color–Word naming*: P is shown the same list of words, but each word (a color name) is printed in a different color ink than the name of the word, e.g., the word RED is printed in any one of the other colors. P is told to name as fast as possible only the *color* of the *ink* in which each word is printed, while ignoring the name indicated by the printed word.

In literate, monolingual adults, the times taken for each of these tasks, from shortest to longest, is typically in the order B, A, C. Various scoring formulas based on the time taken for each task have been suggested by different investigators. Factor analysis reveals that the simplest scoring formulas are the best (Jensen, 1965; Carroll, 1993, pp. 490–492). The color/print interference effect is best measured by C–A ; color naming controlled for reading speed is best measured by C–B; reading speed is best measured by B.

The so-called Stroop effect (i.e., C–A) is remarkably strong, especially among college students. There is an approximately 3:1 time ratio between C and A. Some individuals are so frustrated by the C task requirement that they break down momentarily, while others stammer, stutter, gesticulate, clench their fists, or stamp their feet during this part of the test. Obviously literate persons are unable to ignore the printed words even when they try their best to do so. Having to suppress their implicit response to the printed word makes it surprisingly difficult to utter the intentional response, viz., the actual color of the print.

Because the Stroop test requires P to speak words aloud, there is still a question of whether the Stroop effect would be manifested if administered in a binary CRT format. That is, using a binary response console with an HK (see previous Figure 2.4), P is shown the word of a color name and manually responds *Same* or *Different* according to whether the actual *color* of the printed word (on a computer monitor) is the same as (or different from) the color *name*. The control conditions would measure same–different CRT to colored rectangles and to color names presented in black print. Such an experiment might reveal whether the Stroop effect acts mainly on speech output or is a more central interference phenomenon that can be manifested independent of overt vocalization. Measurement of the Stroop effect in terms of RT is scarce (e.g., Glaser & Dolt, 1977; Koch, Gobell, & Roid, 1999). The latter study, based on a manual response RT, suggests that the Stroop effect is not entirely dependent on vocalized responding.

## Timed Involuntary Responses

### Modified Blink Reflex

The eye blink reflex elicited by an external stimulus such as a loud noise or flash of light or a tap on the forehead is an involuntary response. It is distinguishable in latency and amplitude from spontaneous blinks and blinks made under voluntary or conscious control. The latency and amplitude of the blink reflex, elicited by a mechanical tap to the glabella (i.e., the area between the eyebrows), can be measured noninvasively with great precision by means of an optical apparatus designed for this purpose. The elicited blink latency measurements occur within a range of 20–100 ms and they show highly reliable individual differences. The blink reflex is an involuntary, low-level reaction mediated by brainstem mechanisms and does not seem to reflect the higher brain centers involved in information processing. However, when the eliciting stimulus (a tap on the glabella) occurs in temporal proximity to another stimulus (e.g., a tone) that does not itself elicit a blink, both the latency and amplitude of the blink reflex are either increased or decreased depending on the temporal relationship between the glabellar tap and the modifying tone. The reflexive response thus altered is called the *modified blink reflex* (MBR). Its difference from the unmodified, or baseline, blink reflex involves the higher cerebral centers involved in information processing. Therefore cognitive psychologists have looked to the simplicity and involuntary nature of the MBR for its potential as "a single strategy-free procedure, which will provide a pure measure of speed, or efficiency, in [information] processing across all age groups" (Smyth, Anderson, & Hammond, 1999, p. 14). The same reference describes in detail the precise procedures for measuring the MBR, and provides references to the literature on this subject. In their own study, Smyth et al. (1999) found that MBR showed low to moderate age-partialled correlations with IQ and IT, although there were erratic and puzzling correlational differences between males and females. Although the results leave little doubt that MBR taps central neural processes reflected in cognitive abilities, the nature of its correlations with age and sex would require systematic studies based on much larger samples to allow any firm conclusions.

MBR will remain esoteric until it is clearly shown that there are reliable and replicable correlations of moderate size between MBR variables and certain cognitive or personality variables. Smyth et al. (1999) are optimistic about its potential usefulness in cognitive research: "Because the unmodified blink reflex involves only simple, low-level processing while its modified counterpart can reflect more complex processing carried out in higher structures, blink reflex modification provides a useful paradigm with which to examine the relative contribution of speed at these different levels of processing" (p. 33).

### Critical Flicker Frequency

Research on this phenomenon, called *flicker fusion*, has fascinated a few physiologists and psychologists for over two centuries (Landis, 1953). Its psychological correlations are almost entirely with variables in the realm of neuropathology. So extensively has critical flicker frequency (CFF) been researched that, from the standpoint of physiological optics and experimental psychology, virtually everything that could be known about CFF is seemingly now known. A brief review of the essential facts is provided by Woodworth and Schlosberg (1956, pp. 380–382). When a light flashes at a slow rate, P reports seeing a distinct series of flashes separated by dark intervals. As the flash rate increases, the distinctness of the flashes changes its appearance to that of a continuous *flicker*. A further increase in the flash rate leads finally to the rate known as *flicker-fusion* or the CFF at which there occurs the appearance of a perfectly steady light. These changes normally occur all within a range from about 5–50 Hz (i.e., hertz or cycles per second). Few people realize that an ordinary light bulb does not really emit a continuous light, but a succession of intermittent on/off flashes occurring at a rate that exceeds the upper limit of CFF for humans. A motion picture consists of a series of still pictures, each separated by dark intervals, shown at a constant exposure rate of about 60 Hz. The same is true for television. The fusion threshold, or CFF, varies as a function of light intensity, the relative durations of the light/dark phases within each cycle, the area of the light field, the part of the retina stimulated by the light, and chronological age (CFF gradually decreases with advancing age in adults). Yet even with all these variables strictly controlled, there remain clear-cut individual differences in CFF.

With the proper equipment, using an eyepiece to control ambient light and a monochrome red diode as the variable light source, each eye is tested separately. P operates a dial that controls the flash rate by finely graded steps. In several trials going from a slow to a faster rate, and then from a fast to a slower rate, P repeatedly adjusts the control dial so as to zero in precisely on the rate at which flicker-fusion occurs, i.e., the CFF. Using the best available apparatus, designed for research in physiological optics in the School of Optometry at Berkeley, it was possible within a 10 min test session to measure CFF in university undergraduates with a CFF reliability coefficient of .95 (Jensen, 1983). Data from the same study revealed a correlation of virtually zero between CFF and IQ. A metaanalysis of all previous published studies of the CFF × IQ relationship shows all the correlations to be normally distributed around a mean of zero. Evidently, there is no correlation between intelligence and CFF. Individual differences in CFF seem to be localized entirely in the visual system. Although P has to make subjective judgments of the occurrence of CFF, what is being judged is information delivered by the visual system. Although the accuracy of P's judgment itself may be related to P's IQ, the involuntary objective CFF *per se* is virtually unrelated to P's cognitive ability. This brings into question the potential usefulness of CFF in the study of individual differences in behavior mediated by central processes, cognitive or otherwise.

## Paper and Pencil Chronometry

A number of PP tests have been used chronometrically, and many other possible PP tests could be patterned after the paradigms based on computerized devices. The PP tests invariably measure the overall time taken for responding to a large number of stimulus presentations (i.e., elementary cognitive test items) that would each be measured as single trials in the typical electronically administered RT paradigm. Mean time per item on the PP test, then, is indirectly estimated as *(total time)/(number of items)*. Because the perceptual and motor demands are so different across various PP tests, what is lost is the advantage of the uniformity and precision of measurement afforded by having one and the same response console for different tasks and the timed measurements for single trials. The measurement of RT on single trials also allows the identification and possible elimination of outliers in the distribution of an individual's RTs. Having chronometric data for single trials also makes it possible to obtain measures of intraindividual variability (RTSD) and certain time series statistics as well as permitting the investigator to plot frequency distributions of single RTs, and to rank an individual's RTs by their order of magnitude. In the quantitative analysis of individual differences, these are all critical features of RT data. Therefore, PP tests are generally less desirable in chronometric research than methods, which permit the measurement of RT on single trials.

Chronometric data from PP tests, however, have proved useful in strictly correlational studies in which scale properties and other features of performance besides the total time taken are deemed unimportant. Studies based on PP analogs of computerized RT paradigms also show significant correlations (around + .50) with their electronic counterparts, and the PP and computer versions of the same or similar information-processing paradigms also show comparable correlations with IQ and other nonspeeded cognitive tests (Lindley, Smith, & Thomas, 1988; Neubauer & Bucik, 1996; Neubauer & Knorr, 1998).

# Chapter 3

# Reaction Time as a Function of Experimental Conditions

## Contrasting the Approaches of Experimental and Differential Psychology

Although the main focus of this book is on individual differences in Reaction times (RTs) and their relationship with other psychological variables, investigators in that sphere should be mindful of the many experimental variables that can affect the measurement of RT. To date, the research literature on the differential psychology of RT is minuscule compared to the vast literature on the experimental psychology of RT. Yet, in any RT experiment individual differences generally account for more of the total variance in RT than do the experimental variables being investigated. Experimentalists, however, would argue that the causal nature and meaning of individual differences in RT in any given paradigm cannot begin to be understood independently of some kind of information-processing model that can be explicitly tested only by experimental means. They would claim that without such empirically testable models to account for the relationship between RT and the experimentally manipulated variables, the differential psychologist would only be probing a "black box" for individual differences in RT that causally reflect just anyone's surmise, however impressive may be the correlations of RT with other psychological variables.

Experimentation using RT as the dependent variable is aimed principally at testing hypotheses derived from theories, or models, intended to explain the functional relationships observed between RT and carefully controlled independent variables. The main findings of these relationships, but not the various theoretical models of RT aimed at explaining them, are summarized in this chapter. The explanatory models are typically precise mathematical formulations of hypothetical information processes. These may include hypothesized neural processes as well. For example, a model of the positive relationship between intensity of the reaction stimulus (RS) and the speed of reaction (or 1/RT) could postulate a neural recruitment process — that a more intense RS activates a larger number of independent neural pathways, which increases the probability that some critical threshold of neural activation needed for a response is reached within a certain time period. Another model might postulate a stochastic neural signal/noise ratio — a level of background neural "noise" (i.e., spontaneous random firing of neurons) must be exceeded by the neural excitation elicited by the RS for a response to occur, and the stronger the RS, the greater is the signal/noise ratio. Each model, if precisely formulated, leads to somewhat different predictions, which can be rigorously tested with experimental ingenuity.

Here we will not go into the model-testing aspect of the experimental psychology of RT, except to point out some of the basic experimental variables that research has shown to influence RT. There are two excellent books that deal extensively with the modeling of basic paradigms and experimental variables in RT research, particularly paradigms of simple reaction time (SRT) and choice reaction time (CRT), and they reference most of the theoretical and empirical literature in this field (Luce, 1986; Welford, 1980a).

Probably the most basic distinction, between "peripheral" and "central" processes, regarding the RT phenomena of primary interest to experimental RT theorists, is also a critical consideration for the differential psychology of RT. It is well stated by Duncan Luce (1986):

> The first thing that simple reaction-time data seem to suggest is that the observed reaction times are, at a minimum, the sum of two quite different times. One of these has to do with decision processes, invoked by the central nervous system, aimed at deciding when a signal has been presented. The other has to do with the time it takes signals to be transduced and transmitted to the brain and the time it takes orders issued by the brain to activate the muscles leading to responses. These processes may, in fact, be quite complex and may also be affected by the stimulus conditions. Since these are believed by many to be entirely different processes, we will attempt to decompose them and deal with them separately (p. 94).

### Within-subjects and Between-subjects Designs

This is the most obvious distinction between the experimental and differential approaches to the study of RT. Nearly all experiments are exclusively within subjects (WS) designs; that is, they are focused on the effects of different values of the experimentally manipulated independent variables on variation of the RT *within* a given individual. Such studies can be performed with a single subject, or $N = 1$. Having $N > 1$ is incidental, and is used to increase the generality and statistical significance of the observed effect in a narrowly specified population. In experimental work on RT, the $N$s are typically very small, $N = 1–3$ or 4 is not unusual. But each individual is tested repeatedly, often for many hundreds or even thousands of trials. The advantage of a small $N$ with many trials per subject is that the between subjects (BS) variance is kept relatively small (or 0 for $N = 1$), so that the effects of the experimentally manipulated variables stand out clearly. The WS variability is generally much smaller than the BS variability, and so an error term based entirely on the WS variance is preferable. Simple replications of experimental effects based on small $N$s, rather than the use of large-scale statistical designs, are most typical in experimental RT research. However, if both the WS and BS variances are adequately estimated by using a large $N$ and many repeated trials, it is possible to evaluate their effects separately by the analysis of variance.

Differential RT research, however, is always a BS affair. The conditions of RT measurement are held constant, while individual differences in RT are registered. A large $N$ is necessary, of course, because the full distribution of RT in some specified population is of interest, usually along with the correlations of RT with other individual difference variables, and the reliability of correlations, which is directly related to $\sqrt{N}$, requires a fairly large subject sample for hypothesis testing.

**Observed variables and latent variables** This distinction between observed and latent variables is made by both the experimental and differential approaches, although in somewhat different ways. Science begins with *observed* variables (also called *manifest* variables) in the form of raw observations of phenomena or measures or other quantitative indices achieved by means of various measuring instruments, tests, and the like. All of these constitute the observed variables. Regularities, lawfulness, and similarities in the relationships between observed variables that differ in surface characteristics lead to a hypothesis that the observed relationships among superficially different variables result from their being imperfect or biased representatives of some unobservable, or *latent,* variable, whose nature can only be inferred from either experimentally induced or naturally occurring variation in the observed variables.

Perhaps the most fundamental example of a latent variable is the concept of a *true score* ($t$) as postulated in classical measurement theory. It represents an obtained or observed score ($X$) as the sum of $t$ plus some random error ($e$) deviation from $t$, i.e., $X = t \pm e$. This basic formulation applies to all kinds of empirical measurement in every field of science. When we measure someone's height, for example, we are measuring $X$, not $t$. But if we are really more interested in $t$ than in $X$ *per se*, measurement theory tells us that the larger the number of measurements obtained of $X$, the closer does the mean of all the $X$s approach $t$. The idea that there exists a true value, $t$, to be approximated ever more closely by averaging $X$s is a hypothesis, and $t$ is the latent variable we wish to estimate. Also, when we subject a coefficient of correlation between two observed variables, $X$ and $Y$, to a correction for attenuation (by dividing the correlation by the geometric mean of their reliability coefficients), we are attempting to get a better estimate of the true-score correlation between $X$ and $Y$, i.e., the correlation between $t_X$ and $t_Y$. So it is with modeling RT in experimental psychology. Certain classes of phenomena are explained in terms of inferred properties of the nervous system or of some abstract representation in terms of an information-processing system that possesses whatever hypothetical mechanisms (*information processes*) are inferred to account for the observed variables. Information processes are the latent variables of experimental cognitive psychology, which often uses mental chronometry in its investigations (Posner, 1978; Luce, 1986).

Differential chronometry investigates individual differences in the information processes modeled by experimental chronometry. It also seeks to describe the *structure* of individual differences in these chronometric variables. The concept of structure in this context implies latent variables, here referred to as *factors*, which are hypothesized sources of variance intended to describe or "account for" the correlations between individual differences on various chronometric measurements and correlations between these chronometric measurements and measurements of individual differences obtained from other, nonchronometric behavioral variables. Spearman's g factor is a well-known example of such a latent variable. Three closely related methodologies are used for these types of correlation analyses: components analysis, factor analysis, and structural equation modeling. All of them are well explicated by Loehlin (1998).

## Experimental Variables and Procedures that Affect RT

The effects of the following variables are described here only in general terms of direction and magnitude. Specific quantitative values vary markedly with the nature of experimental conditions and of the subject sample, of which detailed descriptions would necessarily have to accompany any quantitatively precise conclusions. The various experimental conditions that are known to influence RT need to be taken into account in the design of any particular paradigm and apparatus for the measurement of RT used in studies of individual differences.

The aim in differential work is generally to set the experimental parameters of a given paradigm at values that will contribute the smallest ratio of WS variance to BS variance in the RT measures of interest. The variables of central interest will usually be information processes rather than peripheral sensory-motor variables. For example, in a CRT paradigm intended to tap a decision process, the RS should have a level of stimulus intensity such that very little if any of the BS variance in the CRT measures is contributed by individual differences in sensory acuity.

### Preparatory Stimulus and Preparatory Interval

The effects of preparatory stimulus (PS) and preparatory interval (PI) on RT have been studied extensively (156 references in Niemi & Näätänen, 1981). They are a means for achieving some degree of quality control over the obtained RT data. Without a PI of limited duration, a subject's trial-to-trial RTs are more variable; hence their mean is less reliable. The PS serves to alert the subject to the impending RS, while the PI allows a build-up of tension and increasing readiness to execute the required response after the occurrence of the RS.

A high level of discriminability between the PS and RS is essential. It is best assured by having the PS and RS of different sensory modalities, such as auditory and visual, so that the main information processing burden follows the RS stage of the RT task. The pitch and intensity of the auditory PS, up to a point (about 300–600 Hz at 50–60 dB), is positively related to the speed of response to the RS. The PS should be uniform and of sufficient duration to be unequivocal on every trial; durations from 0.5 to 1.0 s are commonly used.

The PI (i.e., the elapsed time between onset of the PS and onset of the RS) can be either *constant* or *varied* across trials. A constant PI has the disadvantage of allowing subjects to improve their anticipation of the occurrence of the RS. Hence, individual differences in the tendency to adopt an anticipation strategy may add a theoretically unwanted source of BS variance to the measurement of RT. The RT task then takes on the characteristic of a time-estimation task, an ability in which there are reliable individual differences. (It is one of the abilities measured by the Seashore Musical Aptitude Test.) By varying the PI randomly (or pseudo-randomly) over trials, the effect of the PI on RS is considerably lessened. Anticipatory responses are minimized and E can be more confident that the RT actually reflects the response to the RS *per se*. PIs of less than 1 s should not be used, as they fail to serve the intended purpose of the PI. The optimal range of PIs is 1–4 s.

Two types of random PIs are used: *stepped* intervals or *continuous* intervals. Some subjects learn to distinguish stepped PIs (e.g., randomly ordered PIs of either 1, 2, or 3 s) and can anticipate the RS when reaching the recognizable last step of the PI. PIs of continuous random lengths prevent the subject from anticipating exactly when the RS will occur, although expectation and readiness naturally mounts steadily during the PI. If stepped PIs are used, it is better that they be presented in a pseudo-random order; for example, allowing not more than three successive repetitions of the same PI to occur throughout the entire sequence of trials.

The duration of the PI has a small but reliable effect on the subsequent RT. It therefore contributes to reaction time standard deviation (RTSD), that is, trial-to-trial variability in RT. A small range of PIs, therefore, is desirable. The PI effect is more pronounced in children than in adults. When the PIs occur in a random or irregular order and vary between 1 and 16 s, the corresponding mean RTs are U-shaped with the shortest RT at PIs of about 2–4 s. The U shape of this function becomes increasingly shallower from early childhood to early adulthood (Elliott, 1970, 1972).

### Modality of the RS

RT to a visual RS is longer than to an auditory RS. For moderate stimulus intensities, the fastest visual and auditory SRTs are about 180 and 140 ms, respectively, but these values may converge as stimulus intensity increases to well above threshold. RT to touch is about 150 ms. Transduction of a visual stimulus takes longer, presumably because it involves a chemical process in the retina, whereas audition involves a quicker, mechanical action. Other senses have been tested as RS; for example, taste, olfaction, and pain. The stimuli for these senses are hard to control experimentally, and the reactions elicited are relatively slow and imprecise. Hence, sensory modalities other than the visual and auditory are seldom if ever used as the RS in research on information processing.

### Intensity, Area, and Duration of the RS

Speed of reaction increases as a function of intensity, area, and duration of the RS. Intensity of the RS has a greater effect on RT, which approaches asymptotic values at stimulus intensities well above threshold. The total stimulus strength is simply a monotonic function of area × intensity. Precise plots of RT as a function of stimulus intensities are shown in Woodworth and Schlosberg (1954, pp. 20–24). The variability of RT is also a function of RS intensity, and there is a constant proportionality of the effect of RS intensity between the mean and standard deviation of RT (Luce, 1986, p. 64); i.e., the coefficient of variation (CV = $\sigma/\mu$) is fairly constant across different levels of stimulus intensity.

A *summation* of stimuli (e.g., binocular versus monocular vision, or binaural versus monaural hearing) increases the total stimulation in the same sensory modality, resulting in quicker RT. Different stimulus modalities, however, do not summate. When the RS consists of simultaneous visual and auditory stimuli, the RT is the same as that for the auditory stimulus alone, as the ear is a speedier stimulus transducer than the eye, and the energies of the visual and auditory signals do not summate in the brain.

In most RT procedures, RS duration is terminated by the subject's RT response. Fixed RS durations shorter than the RT have little effect on RT when they are beyond about 40 ms duration and when the intensity of the RS is well above threshold. In research on individual differences it is well to maintain the effect of stimulus conditions on RT near their asymptotic levels for the vast majority of subjects if the focus of investigation is on central processes.

## Complexity of the RS

This is one of the strongest effects of the RS on RT. Any feature of the task that in the least increases the complexity of the RS, such as going from SRT to CRT or discrimination reaction time (DRT), increases the RT. RT is monotonically related to the number of choice alternatives. Every RT paradigm is a variation on this theme and it is the types of variation in RT resulting from different task demands (represented in the RS) that are of greatest theoretical interest to both experimental and differential psychologists. So much will be said about the complexity variable in later chapters that there is no point in elaborating on the topic here. RS complexity and task mastery gained through prolonged practice have opposite effects on RT. The more complex the RS demands, the greater is the effect of practice in decreasing RT. But even at the asymptotic levels of RT gained through extensive practice, RT will still vary as a function of RS complexity. In CRT the probability of occurrence of a particular RS, when it is not equal for every RS, also affects the RT, a higher probability making for shorter RT. In differential research, probabilities of the different RS alternatives are usually kept uniform.

## Effects of Practice on RT

This topic has been treated extensively in the experimental literature on RT (Teichner & Krebs, 1974; Welford, 1980b). One of the perennial questions in differential research on RT is whether the RT maintains the same factorial composition (i.e., the same pattern of correlations with other variables) at various points along the trajectory from unskilled to highly skilled or mastery performance. Do the slower RTs in early practice trials represent all the same ability components as the faster RTs attained after a great many trials? This has always been problematic for differential research because a marked practice effect indicates the effects of a learning process in addition to the speed of information processing *per se* that may be the variable of primary interest. Also, the presence of practice effects implies an ambiguity in interpreting the performance of subjects who have had different amounts of experience in RT tests. Practice effects are greater as task complexity increases in number of choice alternatives, lesser stimulus discriminability, and degree of *S–R* incompatibility.

   SRT to light, with maximum *S–R* compatibility shows virtually no practice effect (i.e., improvement) in RT after the first 10 trials. For CRTs, however, the practice effect persists over at least 10,000 trials and the CRT is a decreasing linear function of the logarithm ($\log_{10}$) of the number ($N$) of trials. The average slopes of the regression of CRT on $\log_{10} N$ for 2-, 4-, and 8-choice RTs derived from a number of studies were $-0.099$, $-0.169$, and $-0.217$ s (Teichner & Krebs, 1974, p. 82). These figures indicate that after the first 10 trials, the *rate of decrease* in CRT is quite slow and it becomes slower with each successive block of trials. On a 2-choice RT task, for example, if the RT on the first trial is 0.725 s, it is 0.626 s on the 10th trial, 0.527 s on the 100th trial, and 0.428 on the 1000th trial. In differential work, a large number of trials is seldom feasible and so one has to live with the existence of practice effects and determine whatever effect they have for the given procedure and the total number of trials that can feasibly be used. The investigator takes advantage of the fact that the largest practice effects occur in the early trials, and allows a fair number of discountable practice trials (e.g., 20 or so) before beginning the test trials. If it is important for the investigator's purpose, a systematic trend in RT over trials can be measured or removed by regression methods. A generally more important question for the differential psychologist, however, is whether individuals remain in much the same rank order in their mean RT throughout successive blocks of trials. This is essentially a question of RT reliability, which is treated in Chapter 5.

### Sequential Effects on RT

For differential research, the research on sequential effects on RT (reviewed by Kirby, 1980) supports the recommendation that for any given paradigm, every subject in the study sample should receive exactly the same random (or pseudo-random) sequence of RS alternatives. Moreover, every aspect of the RT task should be kept as identical as possible for all subjects. This has the effect of reducing RT variance due to sequential effects, which constitute an unavoidable part of the experimental error in the measurement of individual differences. These sequential effects tend to average out in the mean RT when it based on a large number of trials, but they do not average out in the standard deviation (SD) of each individual's RTs. It is especially important that this error component should be held as constant as possible in studying intraindividual variability in RT, which is measured as the SD of the subject's RT over a given number of trials (RTSD). This measure of RT consistency has become an important variable in differential RT research because of its surprisingly high correlation with IQ.

Assuming that the RS alternatives are equiprobable, the magnitude of sequential effects increases with the number of alternatives. A *repetition* of the same RS decreases RT; a nonrepetition or *alternation* of the RS alternatives increases the RT; and the alternation effect somewhat outweighs the repetition effect.

These sequential effects are partly a function of the interstimulus interval (ISI), i.e., the elapsed time between the onset of the successive RSs. Sequential effects are minimized by ISIs of 1 or 2 s. In most of my own chronometric work, I have usually used a fixed 4 s ISI or an ISI controlled by the subject whose pressing of the home key causes the PS ("beep") to sound after 2 s.

The literature is strangely sparse and inconsistent on the sequential effects of error responses on RT (i.e., making the wrong response to the RS). The "common wisdom" is that the RT is slower for erroneous responses under nonspeeded instructions and is faster under highly speeded instructions (see the section on "speed–accuracy trade-off," below), and an erroneous response on a given trial causes a somewhat slower RT on the very next trial (and possibly to a lesser degree on a few subsequent trials). Different RT paradigms have different average error rates for subjects at a given level of ability, and *E*'s instructions are generally aimed at keeping subjects' error rates as low as possible for a given paradigm and subject sample. (Procedures for dealing with response errors are taken up in Chapter 4.)

### Speed–Accuracy Trade-off

This becomes an important consideration in all forms of choice or discrimination RT, in which there are typically wrong responses at rates upwards of 2 percent, depending on task complexity and subjects' ability level. For any given CRT task, the subject cannot maximize both the *speed* of response and the *accuracy* of response simultaneously. Hence there is said to be a *trade-off* between speed and accuracy. The degree of trade-off depends on task complexity (which affects response error rates), and the relative degrees of effort the subject expends on either speed or accuracy of response. The *E* attempts to control the subject's speed–accuracy set through *E*'s instructions to the subject, which may emphasize one or the other attitude toward the task. Typical instructions are "Respond as quickly as possible without making errors." Error rates are reduced by immediate informative feedback indicating whether the subject's response was right or wrong. This is taken care of automatically by procedures in which the RS is instantly turned "off" by the subject's correct response. Emphasis on accuracy can be pushed further by reminding subjects that errors are counted and subtracted from their overall performance "score." The aim is to keep error rates as uniformly low as possible for all subjects. In any case, a record should be kept of all errors. The various ways of handling errors in RT data analysis depend on the investigator's purpose, as explained in Chapter 4.

The consequence of the speed–accuracy trade-off in regulating response errors depends on whether one is considering a WS experimental design in which reaction speed is manipulated by *E*'s instructions emphasizing either speed or accuracy or a BS differential design in which all subjects receive the same instructions. The relationship between RT and error rate is *negative* in the WS design and it is *positive* in the BS design. That is, an individual who is urged to increase response speed (i.e., decrease RT) tends to sacrifice accuracy (i.e., increases error rate). But among any group of individuals given the same instructions by *E*, those individuals with longer RT also generally have higher error rates. It was once mistakenly conjectured that IQ is negatively correlated with mean RT because individuals with a higher IQ tend to adopt a speed–accuracy trade-off, sacrificing accuracy for greater speed. This proved not to be the case, as massive data showed that higher-IQ individuals have not only faster RT but also greater accuracy in responding. So BS speed and accuracy are positively correlated.

The WS speed–accuracy operating characteristic curve is illustrated graphically in Figure 3.1. The BS relationship between RT and error rate for tasks differing in complexity is shown in Figure 3.2. On the simplest task (e.g., SRT) the RTs of three hypothetical persons A, B, and C are shown as the same, with low error rate and short RT. On a more complex task (e.g., CRT) the persons are shown to differ both in RT and error rate. The speed–accuracy trade-off curves differ for each person; as the RT is reduced, the error rate is increased, but persons maintain their same relative positions, so the correlation between speed and accuracy remains positive at all levels of task complexity.

### *Motivation, Incentives, Payoffs, Informative Feedback*

These conditions, when experimentally manipulated, have remarkably little or no effect on RT and IT measures of information processing speed (studies reviewed by Deary, 2000a, pp. 157–59, 197). Motivation or arousal have been measured physiologically, for example, by involuntary changes in responses controlled by the autonomic nervous system, such as pupillary dilation, which closely reflects the level of a person's effort elicited by a given cognitive task (e.g., a test item) for a given individual. At any particular level of task difficulty (objectively measured as percent of subjects who "pass" the item) the abler, or

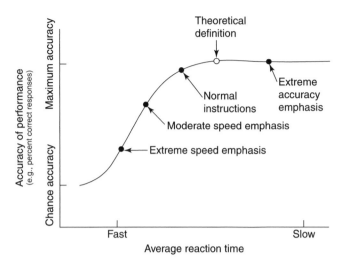

Figure 3.1: An idealized speed–accuracy operating characteristic curve, showing the relationship between an individual's average RT and the accuracy of performance under different speed instructions, and the theoretical definition of an individual's true RT. (Reprinted from Pachella, 1974, with permission of Lawrence Erlbaum Associates.)

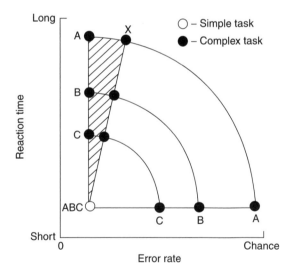

Figure 3.2: The BS relationship of RT to error rate as a function of differences in task complexity. The arcs describe the speed–accuracy trade-off for hypothetical persons A, B, and C, who are shown here as having the same RT on the simplest tasks. The shaded area represents the desirable range of speed–accuracy trade-off. It should be noted that a total absence of errors on any RT task involving choice raises the suspicion that the person's measured RT may be too slow to be an accurate approximation of the true or theoretically defined RT. Hence error rates in the region of 1–5 percent are considered acceptable. (Reprinted from Jensen, 1985, with permission from Plenum Press.)

more intelligent, subjects "pass" the item with lesser mental effort (as indexed by the pupillary response) than is expended by less-able subjects (Ahern & Beatty, 1979). It is not a matter of abler persons trying harder, they simply process information more efficiently. Offering subjects as much as $20 to improve their inspection time (IT) speed, for example, had no beneficial effect. The difficulty level of the tasks typically used in RT studies is so low and the relatively short time taken by the typical test session are such that traits like achievement, motivation, persistence, and level of aspiration apparently have little opportunity to be manifested. Although these personal variables may be crucial for climbing the ladders, beating the competition, and persevering in the face of difficulties involved in long-term or overall lifetime achievements, there is no evidence that they play a significant role in RT performance. These chronometric tasks largely tap an entirely different category of individual difference variables, mainly those involved in the brain's integrity and efficiency of information processing.

**Faking slower RT**    Subjects, of course, cannot voluntarily make appreciably faster responses than they normally make when instructed to react as quickly as they possibly can. Only by engaging in very long-term practice on the same task can the RT be shortened, assuming there is no change in the person's physiological state. RT does not remain constant throughout an individual's diurnal cycle, but increases following meals, and with fatigue, and also varies monotonically with fluctuations in body temperature. But faking slightly slower SRTs than one's normal speed is another matter. The conscious intention to respond more slowly than one would normally do under the usual instructions for speed and accuracy causes subjects to far overshoot the mark, so to speak, making it obvious that they are "faking bad." In informal experiments, university students were asked to try to produce slower SRTs than they would normally produce, aiming to approximate the mean RTs typical of mentally retarded individuals, which are some 300–400 ms longer than those of university students. The students actually produced SRTs that are much longer than the average of retarded individuals. Gradually, however, through practice with informative feedback on every trial, they were able to modulate their SRTs to more closely approximate the slower target RT they were aiming for. This shaping of the RT toward a slower-than-normal speed resembles biofeedback conditioning. Subjects report learning strategies based on subtle feelings that develop through practice with feedback, or by blurring the RS by keeping it in peripheral vision, or by squinting.

But why should there be any initial difficulty at all in faking slow? The answer is probably explained by the interesting discovery made by the neurophysiologist Benjamin Libet (1989). By means of electrodes implanted in the primary somatosensory area of the cerebral cortex of fully conscious patients undergoing brain operations under local anesthesia, Libet was able to measure the time required to attain conscious awareness of an external stimulus (a brief electrical pulse on the skin). The average time was about 500 ms — that is, the electrical brain potential evoked by the external stimulus and immediately registered by the implanted electrode had to recruit enough "neuronal adequacy" during a period of half a second in order to reach the threshold of adequacy for the conscious recognition of the stimulus as reported by the subject.

In other words, a totally unconscious brain process was going on for half a second before the subject had any conscious awareness of being stimulated. But there is a "backward referral" of the stimulus in time, which accounts for the person's subjective impression that the actual occurrence of the stimulus and the experiencing of the sensation are virtually simultaneous. Hence in our studies of SRT, in which normal young subjects typically produce mean RTs between 200 and 300 ms, the subjects are responding some 200–300 ms sooner than they are consciously aware of the occurrence of the RS. As they have no conscious awareness of this lapse of time after the occurrence of the RS, they cannot easily control it, although they can learn to do so through training and practice. All of SRT and some large part of most forms of CRT consist of unconscious brain-processing time that takes place even before the subject is consciously aware that the RS has occurred. The phenomenon of backward referral in time would also account for the common impression of subjects that their RT is very much shorter than their movement time (MT), when in fact just the opposite is true. Even though MT involves moving the hand over a distance of several inches, the MT is much shorter than the RT. Subjects are usually surprised that they cannot make their two-choice CRT as fast as their SRT because they have no subjective impression of the difference between the times for SRT and CRT. Such small time differences apparently are below the threshold of conscious discrimination.

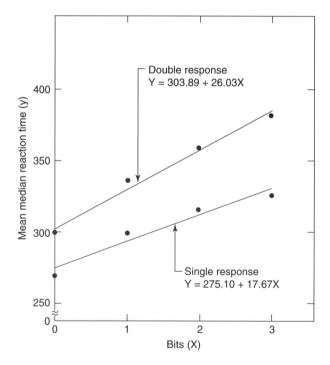

Figure 3.3: The means of the median RTs (in milliseconds) of 25 university students as a function of bits in the Hick paradigm, under conditions of *single response* (lifting finger from home button only) and *double response* (lifting finger from the home button and moving the hand 6 in. to press a button next to the stimulus light, turning it "off"). In both conditions RT is measured as the interval between the onset of the RS and lifting the finger from the home button. (Reprinted from Jensen, 1987b, with permission of Elsevier.)

### Response Characteristics

RT is a function not only of the informational complexity of the RS, but also of the nature of the required motor response. This phenomenon has been investigated by a number of experimental arrangements, such as having different distances through which a subject must move the hand to hit the response key, and by using different sizes of keys that require aiming responses with different degrees of precision (Welford, 1980b, pp. 117–120). In one study, for example, there were two conditions of response: (1) subjects merely had to lift their finger from a key when the RS appeared, or (2) subjects had to lift their finger from a key and grasp a tennis ball hanging by a string. The RT was 20 percent longer under the second condition. This effect of the complexity of the motor response on the RT has become known as "Fitts's Law" after Paul M. Fitts (1954) whose research interest was the psychophysics of motor skills. To determine if Fitts's law might affect the RT measurements obtained on the "Jensen box" apparatus (see Figure 2.10) used to study the Hick phenomenon, a modified procedure was used. Typically, subjects had to make a *double response* when the RS appeared: (1) lift their finger from the home button (i.e., RT), and (2) move their hand 6 in. to press the button next to the RS light (i.e., MT), which turned off the light. When subjects were required to make only a *single response* (lifting their finger from the home button), the typical Hick effect remained, but, in accord with Fitts's law, the mean RTs were shorter by about 25–30 ms, and the linear slope of the linear Hick function was slightly lessened, as shown in Figure 3.3.

The additional 30 ms of RT for the double response is apparently the time taken in programming the specific response to be made when a particular stimulus light goes on. Nearly all subjects keep on pressing the home button until this programming has been done, hence it becomes a part of the RT, not the MT, which remains constant over the range of 0–3 bits (i.e., 1–8 choice response alternatives) in the stimulus display.

Chapter 4

# The Measurement of Chronometric Variables

## Psychophysics, Psychometrics, Chronometrics

For understanding the role of mental chronometry in the behavioral sciences it will be useful to distinguish it from the two other major disciplines in the behavioral sciences concerned with measurement: psychophysics and psychometrics.

*Psychophysics* has a venerable history, probably longer than that of any other branch of empirical psychology, dating from the mid-nineteenth century. Still a field of contemporary research, it deals quantitatively with the functional relationship between the metrical characteristics of various physical stimuli (in every sensory modality) and a person's subjective mental judgments of variation in the quantitative characteristics of these stimuli. A *psychophysical law* is a mathematical formulation of the precise form of the relationship between the magnitude of a stimulus ($S$) measured in physical units and the subjective magnitude of its mental representation ($R$) measured in mental units such as *j.n.d.*s ("just noticeable differences"). An early discovery (1860) of a psychophysical relationship, for example, was Fechner's classic law: $R = k \log S/b,$ where $k$ is a constant of proportionality and $b$ the physical magnitude of $S$ at the absolute threshold of its subjective detection by the subject. This law means that an arithmetic increase in the stimulus magnitude produces a logarithmic increase in the subjective magnitude of the stimulus, or conversely, a geometric increase in physical stimulus magnitude produces an arithmetic increase in the subjective magnitude of the stimulus. Many methods have been devised for measuring or scaling the subjective judgments of differences in physical magnitudes (Guilford, 1954; Stevens, 1975). As the measurements of mental judgments are anchored to the magnitudes of physically measurable stimuli, the validity of the psychophysical methods for deriving mental measurements is tested directly by their mutual concordance and their goodness of fit to the simplest and most generalizable mathematical functions relating the mental to the physical measurements, called the *psychometric function*. These psychometric functions are viewed as probes of the brain, revealing certain of its fundamental operating principles. The measurement problems *per se* of psychophysics seem considerably less formidable and contentious than are those in psychometrics.

*Psychometrics* is virtually synonymous with what is known as *test theory*, or the construction of mental tests composed of *items* to each of which the subject makes a response that is scored quantitatively (e.g., 0, 1, . . . , $n$) according to an explicit set of rules or a scoring key. The total raw score on the test as a whole is simply the sum of the item scores. Psychometrics consists of methods for (1) generating and selecting items to achieve certain desired characteristics of the whole test, (2) "norming" (i.e., determining the frequency distribution) the total test scores for a specified population, (3) standardizing and transforming the raw scores to achieve a statistically meaningful and practically convenient scale (e.g., IQ with population mean = 100 and *standard deviation* (SD) = 15), (4) determining the *internal consistency reliability* of the tests based on interitem correlations and item-total score correlations, (5) determining the *stability* of test scores by test–retest and equivalent forms correlation), and (6) establishing the external *predictive validity* of the test by showing its correlation with some variables outside the test domain (also called *criterion validity*), or (7) empirically testing its *construct validity* by showing that the test's correlation (or absence of correlation) with certain other variables, properties, attributes, or outcomes are predicted by a theory of the latent trait that the test is supposed to measure. Factor analysis is also a psychometric method, typically used to analyze the covariance structure of a variety of tests to determine the number and characteristics of the linearly independent components of variance among the various tests.

The greatest problem in psychometrics is the measurement properties of test scores, even when all of the most sophisticated methods of test construction have been rigorously applied. The bothersome problem is the uncertain nature of the relationship between the metric of the obtained test scores (however standardized or transformed, e.g., IQ) and the metric of the *latent trait* the test is assumed to measure. But it is important to emphasize that this is a wholly unimportant concern for virtually all of the practical uses of psychological tests. It is an utterly trivial basis for criticism of tests as they are most commonly used and valued for their practical predictive validity. In these practical uses there is no interest in what the test "really" measures in terms of latent traits, so long as the test scores have a sufficiently large correlation with some criterion of importance to the test user, such as predicting scholastic performance, success in job training, college graduation, occupational status, response to psychotherapy, emotional stability under stress, and the like. Most professionally constructed and published psychological tests and the uses made of them are intended for practical purposes, and their validity for these purposes is empirically testable and generally known for a given test. Psychometrics in this practical sense is an applied technology rather than a theory-based science. It is when we try to use conventional psychometric tests for scientific purposes that they become problematic.

The basic problems can be understood in terms of essentially three types of measurement scales: *ordinal, interval,* and *ratio.* They differ in the amounts of information they convey, which limits the kinds of conclusions they can support.

An *ordinal* scale simply ranks a number of observations (or individuals) according to some property (e.g., total score, or number of correct answers on a given test). Using letters of the alphabet to represent observations, an ordinal (or rank order) scale permits only the conclusions that a < b < c < d; a < c, a < d; b < c, b < d, etc. An ordinal scale does not allow meaningful arithmetic operations on the ranks, such as the inference that $(b - a) + (c - b) = c - a$, for whatever dimension, attribute, or trait the scale purports to represent. Because the magnitudes of the intervals between the ranked items are unknown, it is also impossible to infer any ratios between them, such as a/b = b/d. Because of these arithmetic limitations, some measurement theorists have argued that ordinal scales do not qualify as true measurement. Even though ordinal scale values do not allow meaningful arithmetic operations besides "greater than" or "less than" (symbolized by > and <), ordinal scales are useful when only rank information is needed, such as selecting the top *n* job applicants from a larger applicant pool, in which case the external validity of the rank order is the only important consideration. A physical example of an ordinal scale is the Mohs' scale for the hardness of gems; they are ranked according to which mineral can scratch another mineral — a unidimensional rank scale of hardness going from the hardest (diamond) to the softest (talc). Note that this scale has no true zero point. Statistics such as the mean and SD based on an ordinal scale are essentially meaningless; they convey no information because the mean and SD are perfectly predictable simply by knowing the total number of ranked items and the number of tied ranks. However, ordinal scales for different attributes can be correlated with one another, either by the Pearson $r$ or Spearman's $\rho$ ($r$ and $\rho$ are identical for ranked scores).

An *interval* scale has not only the rank-order property of the ordinal scale but also the property of equal intervals, which permits additive operations with the scores, such as $a - b = c - d$. But because an interval scale has no natural origin or true zero point, multiplicative operations such as ratios between scores are meaningless. The Celsius and Fahrenheit thermometers are physical examples of an interval scale, as their zero points are purely arbitrary. The Celsius scale defines 0 and 100 °C, respectively, as the freezing and boiling points of water. But because the Celsius scale does not define a true or natural zero point, ratios of the scale values are meaningless; 100 °C is not twice the amount of heat as 50 °C, but only 1.15 times the amount, because the true or absolute zero point of temperature ($0°$ on the Kelvin scale) is $-273$ °C. Means, SDs, and all forms of correlation are meaningful statistics when used with an interval scale. Without equal intervals and a true zero point, however, the coefficient of variation (CV $= \sigma/\mu$) is meaningless. The CV is a theoretically useful statistic; provided variables are measured on a *ratio scale*, the CV answers questions such as: are 10-year-old children more variable in *height* or in *weight*? But there is no way that we can answer the same kind of question regarding *spelling* and *arithmetic*.

A ratio scale is an equal interval scale that has a true zero point. The zero point is precisely no more and no less than none of the attribute measured by the scale. A scale that has a true zero point but not equal intervals cannot be a ratio scale. The units of measurement in a ratio scale are perfectly equal throughout the entire scale and are therefore directly comparable for different entities measured on the same scale; the weight of a mouse can be directly compared with that of an elephant, either in terms of a difference or a ratio. All mathematical and statistical operations *per se* are meaningful on a ratio scale, which includes most physical measurements such as length, mass, and time. All linear transformations of ratio scales are of course also ratio scales, and nonlinear transformations have precisely the same meaning for all ratio scales. Moreover, values from a given ratio scale can be meaningfully combined or entered into various mathematical relationships with other ratio scales and the resultant figures will maintain their ratio scale properties. Complex measurements derived from mathematical relationships between more basic ratio-scale measurements, such as volume, density, pressure, and speed, are also ratio scales.

All psychological tests based on items or performance ratings of any kind are essentially ordinal scales. They are often treated as if they were interval scales. But treating what is essentially an ordinal scale as an interval scale wholly depends upon certain assumptions and the plausibility (seldom a proof) of these assumptions. A good example are tests that attempt to measure individual differences in general mental ability. All such tests begin as ordinal scales. To make the raw scores (e.g., number of right answers) interpretable, they are standardized on some reference group (e.g., a random sample of individuals from some clearly defined population). Within this sample each individual's raw score is transformed to a $z$ score $[z = (X - \bar{X})/\text{SD}]$. As the $z$ scores have negative and positive values, they are often transformed to a scale with a different mean and SD to make all values positive: such as the $T$ scale, or IQ, or SAT scores ($T = 10z + 50$); IQ $= 15z + 100$; (SAT-Verbal $= 100z + 500$). It is important to note that the only information conveyed by any one of these transformed scores beyond what is contained in the raw scores is an indication of where the individual stands (i.e., how many SD units from the mean) in relation to the standardization sample (or "norms"). But these scores, like the raw scores from which they were derived, can tell us nothing about the intervals between scores or the location of the true zero point of the trait that the test attempts to measure. The standardized scores are still just an ordinal scale. The true form of the distribution of the trait represented by scores that are really no more than ordinal is unknown and essentially unknowable.

To elevate such scores to the status of an interval scale requires making the critical assumption that the trait in question has in reality a particular kind of distribution in the population of interest. The form of distribution usually assumed, particularly for mental abilities, is that the true population distribution is *normal*, or Gaussian. In the first place, the normal curve (or "bell curve") has great statistical advantages of a strictly mathematical kind, which alone is a powerful argument for turning an ordinal scale into an interval scale, even without further justification. Unless there is a compelling argument that a trait does *not* have an approximately normal distribution, it is statistically advantageous to assume that its population distribution is normal.

One can argue the plausibility that the measured trait is normally distributed by analogy to a number of physical traits that are measured on a ratio scale, such as height, blood pressure, and brain weight, and their population distributions are all approximately normal. Also one can invoke a theoretical argument for assuming normality of complexly determined traits. It is the Central Limit Theorem, a mathematical proof that as individual measurements each represent the sum of an increasing number of small random effects (as may be hypothesized for genetic and environmental influences on mental development) they tend toward a normal distribution (Jensen, 1998b, pp. 101–103).

To produce a normal distribution of test scores, psychometricians resort to either one or a combination of two procedures.

(1) *Item selection* is the primary method by which the form of the raw score distribution is altered by manipulating the difficulty levels (percent passing) of the selected items. Item difficulty levels, for example, can control the symmetry of the distribution about the mean or median and the range of ability over which the items discriminate; further approximations to normality (e.g., controlling kurtosis) can be achieved by selecting items on the basis of their interitem correlations in the population of interest. With enough items to select from, the raw score distribution can be made to approximate a normal distribution in the population of interest.

(2) *Normalization* of scores is a much easier way to achieve normality. It is often used as a final refinement of the approximately normal distribution achieved through item selection. It consists simply of determining the percentile ranks of the raw scores in the population sample and converting these to their normal $z$ equivalents, which can then be transformed to a scale with any desired mean and SD. If we accept the plausible supposition that the trait

being measured is normally distributed, then, of course, normalized scores may be regarded as an interval scale, while recognizing that there is no compelling proof of this.

Elevating such a mathematically devised interval scale to a ratio scale is far more problematic, as it involves further assumptions and extrapolations to locate the scale's true zero point. Where on the IQ scale, for example, is the true zero point of intelligence? The true zero of temperature on the Kelvin scale was discovered theoretically from the kinetic theory of heat, as the point at which all molecular motion ceases. We have no psychological theory of intelligence that predicts the level of ability at which there is no intelligence. The best try that has been made to determine the zero point of IQ was made by one of the most eminent psychometricians, Louis L. Thurstone (1887–1955). He noted that when normalized raw scores were plotted as a function of children's chronological age, the growth curve of average test scores increased in a regular, negatively accelerated fashion. Similarly, the SDs of the scores increased regularly as a function of age. Assuming the mental growth curve is plotted on an interval scale as a function of age, it is possible to extrapolate the curve downward below the earliest age that children can be tested until the extrapolation reaches the age at which the SD of the scores is zero. The true zero of intelligence is plausibly assumed to be the point at which there is no measurable variation in the trait. According to this method of downward extrapolation, the point of zero intelligence is reached in the later months of gestation. On this supposed ratio scale, a neonate has about one-tenth of a 10-year-old's intelligence; a 10-year old has about two-thirds of the intelligence of a 20-year-old. This Thurstonian scale of mental growth is of little interest today because the tenability of the extrapolation is based on two questionable assumptions: (1) that there is an interval scale across the testable age range and (2) that the measured trait is unidimensional throughout the age range.

A measurement methodology based on a set of mathematical models of the relationship between item difficulty and ability, known as *item response theory* (IRT), also called *latent trait theory*, attempts to overcome the main limitations of classical test theory in which the quantitative interpretation of multiitem test scores is entirely dependent on the ability levels encompassed by a particular reference group. In IRT individual differences in ability and item difficulty are represented on one and the same scale, which is not anchored to a particular reference group. Whereas the total score on a conventional test is the number of correct responses, in IRT constructed tests the person's total score is, in effect, an optimally weighted combination of the right and wrong answers on the separate items, the weights being based on the parameters of the item characteristic curve (ICC) for each item. Hence, the number of possible scores obtained from an IRT-scaled test can exceed the total number of items composing the test. The ICC is the function relating the probability ($p$) of passing the item to the ability level of the individuals. The mathematical logic of the resulting IRT scale of test scores derived from logarithmic transformations of the item $p/(1 - p)$ values can qualify as an interval scale provided that the particular IRT model closely fits the item data. The IRT scale has at best an arbitrary rather than an absolute zero point and therefore is not a true ratio scale. The construction of IRT scales typically requires precise item selection from a very large item pool in addition to a huge subject sample, generally calling for an $N$ of between $10^3$ and $10^4$ in order to obtain reliable estimates of item difficulty that fit the given model throughout the full range of ability measured by the resultant test.

For most practical purposes, IRT tests have little advantage over most tests based on classical test theory and factor analysis. The true advantage of IRT is most evident, however, in constructing equivalent forms of a test, and even more so in adaptive testing administered by computer programs. Each person's level of ability can be zeroed in on quite efficiently and then an optimum number of items at the most appropriate level of difficulty for the particular individual can be presented to reveal precisely where the person falls on the common difficulty/ability scale created by the IRT methodology (Weiss, 1995). Thus, people need not spend time on many items that are either too easy or too difficult for testing their level of ability. Therefore, as compared to ordinary tests, the greater efficiency of adaptive testing with IRT-scaled items permits reliable measurement of a number of different abilities or traits in a relatively short time. But the difficulty of constructing IRT tests and the fact that they are not a ratio scale with a true zero means that they do not really solve the main measurement problem of classical test theory.

### Unanswerable Questions with Psychometric Tests

There are certain kinds of questions of practical or theoretical interest that cannot be answered by a psychological test scale that does not have unequivocally additive units of measurement (interval scale) or a natural zero point and hence meaningful ratios between different points on the scale of measurement (ratio scale). This is true even if only some small segment of the theoretically entire scale is being used. The true zero of temperature, for example, is far distant from the range of temperatures that most people have to deal with, yet ratios between any two temperatures are meaningless without reference to the Kelvin scale of temperature with its absolute zero at 273 °C below the freezing point of water, or 0 °C. Here are some questions for which the answers depend on at least interval or ratio scale measurements.

- The form of a particular frequency distribution is accidental or arbitrary and hence conveys no real scientific meaning unless it is unequivocally at least an interval scale.
- The shape of a growth curve and the ratios of points on such a curve which define rate of growth in different segments of the curve requires a ratio scale. This is one of the problems that has long bedeviled developmental psychology in attempts to study the relationship of increments of mental growth to increments in age or in physical measurements. Measures of both variables (not just age) must be a ratio scale. Ordinary tests, such as used for IQ assessment, cannot do the job.
- Profiles of subtest scores, as commonly seen in such tests as the Wechsler Scales of Intelligence and many other similarly constructed tests are comparable only in a norm-referenced sense, each subtest's standard score indicating the individuals ordinal standing in comparison with the standardization sample on that particular subtest. There is no underlying common scale on which these various subtests are directly comparable in the same way that we can compare (by difference or ratio), for example, the heaviest weights a person can lift with each hand.
- Change scores, measuring gain or loss from one point in time to another, require at least an interval scale to be meaningful for any given individual. Individual change scores assume equal intervals. (This assumption is most questionable if scores are based wholly

or partly on different sets of test items, which may not have the same discriminability.) A change score is necessarily a difference between two scores, say $X_1$ and $X_2$, obtained at different times, each of which has some measurement error, so the error is amplified in their absolute difference $|X_1 - X_2|$.[1] If the coefficient of correlation between $X_1$ and $X_2$ is as large as the average of the scores' reliability coefficients, the reliability of the difference scores, $|X_1 - X_2|$, is zero. Since test score reliability in a single test session is usually based on the interitem correlations (i.e., internal consistency reliability), high reliability depends on the tests being composed of very homogeneous items. For the difference score to be meaningful the average reliability of $X_1$ and $X_2$ must greatly exceed the correlation between $X_1$ and $X_2$. This condition is often hard to achieve in the case of mental ability tests unless there is quite a long time interval over which change is measured (thereby lowering the test–retest correlation). Besides changes in normal maturational growth in abilities during childhood, there are insidious mental changes with aging in late maturity and with the psychological effects of various drugs. Now that specific genes or sections of DNA affecting IQ are being discovered, investigators will want to directly compare different genes on the magnitudes of their effects, and this requires mental measurements having, at least, equal intervals.

## Metrical Properties of Chronometric Variables

The metrical characteristics of the main types of data derived from most chronometric tasks are described below, both as to the methods for measuring the essential variables and their typical functions in research on individual differences. Measurements of *reaction time* (RT) and *movement time* (MT) are considered under five main categories: (1) *central tendency*, (2) *intraindividual variability*, (3) *skew*, (4) *slope*, and (5) *accuracy* (or error rate). This discussion is entirely generic, in the sense that no particular aspects of RT apparatus or computer hardware or software are specified beyond the assumed basic elements needed for obtaining measurements: a *stimulus display* unit (e.g., a computer monitor) and a *response console* consisting of a *home key* (HK) and one or more *response keys* (RKs) or push buttons (e.g., Figures 2.4 and 2.10).

All of the raw measurements obtained from the standard RT paradigms are ratio scales, with a true or absolute zero point and equal intervals. These are the fundamental measurements that are the most valuable in chronometric research. They should always be recorded in any experiment. Possible combined measurements, transformations, or index scores derived from the raw measurements should be generally eschewed unless they can be theoretically justified in a particular study. The most common violation of this rule often results from a naive assumption that some *derived score* — a combination or mathematical manipulation of the raw data — better reveals some hypothetical or conceptual component of cognitive activity than do the original measurements. Such a derived score is not really a measurement of an assumed process but is merely a hypothesis that needs to be justified by empirical tests. The raw trial-to-trial measurements are initially best viewed as elements of a distribution whose descriptive statistics are a purely nontheoretical account of a subject's performance on a given occasion. The conventional unit of time measurement in chronometric research is the millisecond (ms), or $10^{-3}$ second (s). It is now routine for these basic measurements and their distributional statistics to be computer-programmed for computation, storage, and printout at the conclusion of each subject's test session.

### Reaction Time

The interval of time between the onset of the reaction stimulus (RS) and P's releasing the HK. A release of the HK (i.e., lifting one's index finger) is usually faster and easier than depressing a key, and it facilitates one's ballistic response to touch the RK, which is cued by the RS. Curiously, according to quantum theory the absolute zero of RT is $10^{-43}$ s, which is a quantum of time, theoretically the shortest possible time interval. As a single-trial RT typically has a reliability of less than 0.1, it is important to obtain a fairly large number ($n$) of trials. In practice, $n$ is determined by the level of reliability needed for E's purpose, and by consideration of the typical subject's capacity for sustained attention and liability to fatigue during the test session. For example, a higher level of reliability for individuals' measurements is needed in a correlational or factor analytic study than in a study based on average group differences, since unreliability, or measurement error, attenuates correlations but tends to "average out" in group means.

An essential condition for RT measurement is the use of preliminary *practice trials*. The $n$ for practice trials depends on the complexity of the task. Following E's oral or printed instructions explaining the task requirements, the $n$ practice trials — always closely observed by E — have two purposes: first, to ensure that the subject fully understands the task requirements and performs accordingly, and second, to allow the subject to gain easy familiarity with the apparatus and acquire facility in performing the mechanical aspects of the task. In most studies it is seldom feasible, nor is it usually necessary, to bring subjects to their asymptotic levels of performance before beginning the test trials. Asymptotic performance is typically reached only after hundreds or thousands of trials. After their first 10 or 20 practice trials, most adult subjects show very small improvements in RT performance in subsequent trials and individual differences in RT show little interaction with $n$. That is, individual differences in the learning curves for RT performance, though differing in overall level, remain fairly parallel across trials. The rank order of individual differences in asymptotic RT differs little from their rank order after the first 20 or 30 test trials. The largest fluctuations in RT occur *between* test sessions rather than *within* a given test session. Most of the average difference in an individual's performance between sessions is probably related to daily fluctuations in physiological state.

Obtaining an average value of the RTs measured over $n$ trials has been accomplished by several methods. The most frequently used are the *mean* RT (RTm) and the *median* RT (RTmd). The common rationale for RTmd is based on the fact that the frequency distribution of single-trial RTs is skewed to the right and the median value (i.e., the 50th percentile, or the point that divides the total distribution of RT values exactly in half) is considered to be the best measure of central tendency of a skewed distribution. Another possible advantage of RTmd is that, unlike RTm, it is hardly affected by outliers, so that, when $n$ is relatively small, RTmd is a more reliable measure of central tendency. Without outliers in the distribution of RTs, the standard error of the mean is smaller (by about 20 percent) than for the median. RT data obtained in my lab shows slightly higher reliability coefficients for

RTmd than for RTm, although the difference in their reliability decreases with increasing $n$. RTmd is also somewhat less correlated with the SD of RTs over trials (RTSD) than is RTm.

But even in view of these desirable features of RTmd, there is still a stronger reason for preferring the mean, RTm. The reason is that means are additive, while medians are not. The consequence, for example, is that if we wish to calculate the average RT of a given subject for whom we have the RTmd on $n$ trials given on each of three occasions, the mean of the medians over occasions will not be equal to the RTmd over all $3n$ trials. On the other hand, the mean of two or more means (each based on the same $n$) is always equal to the mean of the total $n$. Of course, the same applies to group means. (Unlike the skewed distribution of RTs for a single subject, however, the distribution of the mean RTs based on each of some large number of subjects approximates a normal distribution.) Moreover, the mean has a smaller standard error than the median. Therefore, when RT data are subjected to various statistical analyses and metaanalyses, there is greater quantitative precision and less unaccountable "slippage," so to speak, with RT data based on individuals' mean RTs than that based on individuals' median RTs.

Instead of using the RTmd to minimize the effects of outliers in a subject's RT distribution, it is better to use RTm after removing whatever decrement in RTm reliability is caused by outliers. This is accomplished by "trimming" (also termed "truncating" or "censoring") an individual's RT distribution obtained over $n$ trials. A distribution of RTs based $n$ trials is assumed to be generated by some "parent" mechanism or system. Outliers, or values that fall outside this "parent" distribution, are considered members of some foreign distribution resulting from causes irrelevant to E's interest. Identifying and eliminating outliers calls for an explicit rule uniformly applied to all subjects in a given experiment. Statisticians have proposed several methods for "trimming" a distribution's outliers (Kendall & Stuart, 1973, pp. 546–550). In the case of RT, two different "trimming" rules must be followed, one for the lower end of the RT distribution (i.e., excessively short RTs), the other for the *high* end (i.e., excessively long RTs). Both rules can be computer-programmed, so the subject's trimmed RT data are immediately available at the conclusion of testing. The statistics of the trimmed data are automatically calculated by the computer program after the final trial and additional trials are administered to bring each subject's number of trials up to the same total number of "good" trials for every subject in the experiment. The RS for each of the discarded trials is repeated, so all subjects will have the same number of "good" (i.e., "parent" generated) trials for each RS condition. The number of discarded trials is also recorded for possible use in data analysis.

At the *low* end of the RT scale, for all subjects, any RT less than 150 ms is eliminated. An RT of 150 ms is near the physiological limit of RT in humans. RTs of less than 150 ms are extremely rare in typical simple reaction time (SRT) experiments with college students. Although somewhat shorter RTs (around 100 ms) have been reported, they are nearly all the shortest SRTs obtained with extremely intense auditory or visual stimuli, or they occur only after extremely extensive practice over hundreds or thousands of trials.

Generally, RTs shorter than 150 ms result from an *anticipatory response*. That is, the individual's response is already implicitly in progress before the onset of the RS. If such an anticipatory response is a bit too early, a "false alarm" or error is recorded. Although the computerized RT apparatus is designed to short-circuit any response that occurs before the onset of the RS, an RT of less than 150 ms is taken to indicate that the subject's response was anticipatory and not an authentic response to the RS *per se*. Such anticipatory RTs are therefore discarded.

At the high end of the RT scale, the trimming rule for single trials we have found most satisfactory discards any RT which exceeds $3(X - \overline{X})/SD$, where $X$ is the individual's RT on any particular trial, is the individual's RTm based on all trials on which RT $\geqslant 150$ ms, and SD is their standard deviation. This is a part of the computer program for the given RT paradigm.

Speed of reaction is a muddled or improper measure for describing the phenomena that RT is intended to represent. *Reaction speed* is sometimes represented as the reciprocal of RT, i.e. 1/RT. But this is a meaningless and improper measure, because speed is a measure of rate, which involves two distinct dimensions. It is meaningful, at least in principle, to express the speed of information processing in units of *bits per second* (as in Hick's paradigm). But 1/RT *per se* is not a measure of rate (or speed) and cannot be the direct opposite of RT. The two sets of values (RT and 1/RT) based on the same data are always negatively correlated, but they are often far from perfectly correlated, except for rank correlation. Consequently, their Pearsonian correlation with an external variable (e.g., age, IQ) will not have the same absolute magnitude as the rank correlation. RT itself is typically correlated *negatively* with many other variables. In factor analysis the variables are conventionally "reflected" so that higher scores consistently represent better performance. Using the reciprocal of RT to achieve this purpose is clearly a mistake and will yield different results than if the untransformed RTs had been entered into the analysis. The factor loadings for the RT variables can be reversed in sign, noting that the positive loadings for RT variables simply mean *shorter* (not speedier) reaction time. Using the reciprocal transformation of RT can also lead to other statistical inconsistencies with results based on RT itself, such as in the interaction terms in ANOVA. But these need not be elaborated here, as there is never any need to use "speed of reaction" in place of "reaction time" in chronometric research, except when legitimately used in a ratio with another metric variable as a measure of rate (e.g., bits/s).

Transformations of the RT measured in milliseconds are rarely called for in research on individual differences. The advantageous ratio and equal interval properties of RT are lost in some transformations, as is the advantage of having an International Standard Unit of time measurement, which is precisely calibrated by the earth's daily rotation on its axis and the transit of certain fixed stars across the hairline of a telescope in the Greenwich Observatory. The practice of using a square root or logarithmic transformation to overcome the skewness of an individual's RT frequency distribution should be eschewed. In strictly experimental studies of the functional relationship between RT and an *independent* variable (e.g., number of response alternatives, stimulus intensity, amount of practice) it is sometimes advantageous to transform the *independent* variable, usually to a logarithmic scale.

### Intraindividual Variability of RT (RTSD)

Trial-to-trial variability in RT (henceforth labeled RTSD) is an intraindividual variable in which there are large and reliable differences *between* individuals. Moreover, it shows consistent (i.e., correlated) individual differences over a wide variety of RT paradigms and therefore qualifies as a factor or trait. This fact alone clearly distinguishes RTSD from random errors of measurement. It is measured as the SD of RT over some specified *n* trials of trimmed RT data. Unfortunately, RTSD has been rather neglected in chronometric research. Experimental psychologists have often treated it as merely an error component in their RT data, although recent modeling of response times necessarily includes consideration of RTSD as well as higher moments of the RT distribution. It is also noted (e.g., Luce, 1986) that the coefficient of variation (CV=SD/mean) is fairly constant across a wide range of SRTs, implying a high degree of linear relationship between RTm and RTSD. Indeed, there is typically a high-Pearsonian correlation between these variables, which might suggest that one or the other variable is wholly redundant.

*However,* there is good evidence, though perhaps not yet compelling, that RTm and RTSD are not redundant measures of one and the same latent individual differences variable, though they are highly correlated. RTm and RTSD are less correlated, however, if the correlation is calculated between experimentally independent sets of RT data, such as odd and even trials, which eliminates correlated measurement error. Also, the RT and RTSD behave quite differently in too many ways to make it seem likely that they are totally nonindependent variables. In the Hick paradigm, for example, whereas RTm (or RTmd) shows a perfectly linear relationship to bits, or the binary logarithm of the number of RS alternatives), RTSD shows a perfectly linear relationship to the *number* of alternatives *per se* and is exponentially related to bits, as shown in Figure 4.1. Also, RT and RTSD show different (even opposite) interactions with group variables such as age, sex, and race (Jensen, 1992). Another striking fact is that RT (both RTm and RTmd) shows somewhat lower correlations with IQ than does RTSD, despite the lower reliability of RTSD. These issues and their related evidence are more fully discussed elsewhere (Jensen, 1992; Baumeister, 1998). It is agreed that further research on the nature of RTSD is warranted, especially in

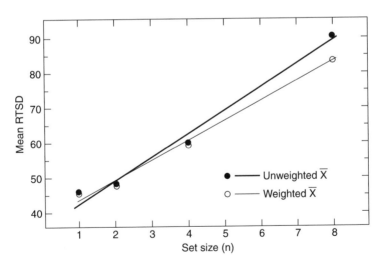

Figure 4.1: Mean SDRT as a function of set size in the Hick paradigm in 18 independent samples totaling 1402 subjects. Both the unweighted and *N*-weighted means are shown; each is correlated .99 with set size (*n*). (From Jensen, 1987a, p. 137. Reprinted with permission from Elsevier.)

view of its significant correlation with IQ independently of RT. The latent trait reflected by RTSD has been conceptualized as "neural noise" in the information processing system. It is also conceived as a negative reflection of "neural efficiency" or "information processing efficiency," similar to Clark Hull's (1943) theoretical construct of oscillation in reaction potential, $_sO_R$. All such constructs, of course, need to be empirically tested for their usefulness in developing a theory of individual differences in RT and its relationship to other behavioral and neurological variables.

The basic measurement of RTSD should be confined to RTs obtained in a single-test session for a single paradigm and a single set of experimental conditions and procedures, otherwise RTSD is confounded with mean differences between conditions, between tasks, or between occasions. In the Hick paradigm, for example, RTSD should be obtained separately for each level of task complexity (i.e., number of choices or bits). Because SDs are not additive, if an individual's average RTSD over different levels of task complexity or different tasks is calculated, the values of RTSD should be converted to variances (i.e., RTSD squared) and then as the root mean square, i.e.,

$$\text{Mean SD} = \sqrt{\frac{(\text{SD}_1)^2 + \cdots + (\text{SD}_n)^2]}{n}}.$$

A theoretically important question about intraindividual variability in RT is whether the variation in an individual's RTs over $n$ trials appears to have resulted from a random generator, although there are consistent differences in the parameters of the RT distributions produced by various individuals. It is as if each individual has a store of RTs with unique parameters (e.g., RTm, RTmd, RTSD) for performing a given chronometric task on a particular occasion, and the testing procedure over $n$ trials taps a random sample of RTs from this store. Certain statistics, such as serial correlation, applied to typical RT data suggest that RTs *within* individuals on a single test in which there are anywhere from 15 to 60 trials (Jensen, 1982, pp. 104–105) do not differ significantly from random expectation. Such randomness does not exist for RT statistics (i.e., the moments of an individual's RT distribution) across test sessions one day apart. The within-session RTs appear random, but an individual's RT statistics vary systematically between-days, as if the parameters of the individual's "random generator" vary systematically from day to day. These parameter differences, however, are much smaller within individuals than between individuals, and therefore differences in RT parameters remain highly reliable between individuals despite day-to-day (and even hour-to-hour) fluctuations of RT within individuals. RT is remarkably sensitive to normal diurnal changes in physiological state that occur in body temperature.

As yet, definitive conclusions about the possible randomness of intraindividual variation in RTs cannot be claimed. One argument that would seem to bring randomness into question is the fact that individuals are unable to consciously generate a series of random numbers or intentionally make any kinds of choices that are actually random. This is true even for statisticians who fully understand the concept of randomness. The conscious mind's intentional attempts to produce randomness invariably fail statistical tests of randomness (Kendall & Stuart, 1977, pp. 222–225). However, it has been noted that RTs, particularly for SRT, are often shorter than the time needed for conscious awareness of an external stimulus, and it could well be that randomness of RTs results from a totally involuntary and unconscious periodicity of excitatory and inhibitory states of the central nervous system (CNS) that are not synchronous with the appearances of the RS. Although the parameters of an individual's RT distribution are determined by certain properties of that individual's CNS, the RTs on each trial could be random, not because of any inherent randomness in the CNS (although that cannot be ruled out), but because the regular periodic fluctuations of the CNS at any given moment are not in synchrony with the randomly timed appearances of the RS. Hence, the existence of reliable individual differences in RTSD could indicate stable characteristics of neural periodicity while an individual's random variation in RT merely reflects the asynchrony of the stable periodicity of the individual's nervous system with the experimentally randomized RS intervals. In fact, there is some evidence that with nonrandom or fixed intervals for the onset of the RS, the variability of an individual's RTs over trials is nonrandom but fluctuates in a systematic way in relation to reactions or errors on preceding trials, as shown by significant serial correlations of .10 to .15 between adjacent trials (Laming, 1988).

The *mean square successive difference (MSSD)*, or its square root, developed by the mathematician John von Neumann (1941), is another method for measuring RT variability. The chance probabilities of what is known as the *von Neumann ratio* (i.e., MSSD/Variance) have been tabled by Hart (1942). It is related to the *serial correlation (R)* across trials, and, when $n$ is large, it provides a strong test of randomness; its expected value is $2(1-R)$, and $R \rightarrow 0$ for perfectly random sequences. Although MSSD is used in economics for time series analysis and in physiology (Leiderman & Shapiro, 1962), as far as I know, it has never been applied in chronometric research. However, its use could be a valuable supplement to RTSD, because MMSD measures the variability of sequential values without reference to their deviations from the mean of all the values in the sequence. Also, it does not reflect variance associated with any systematic trend in the sequence of RT values, as may occur over prolonged trials in a RT task in which RT decreases over trials. (Such sources of variation, when significant, should be measured independently.) The MSSD is symbolized as delta squared ($\delta^2$) and is defined as $\delta^2 = [\Sigma(X_i - X_{i+1})^2]/(n - 1)$, where $X_i$ and $X_{i+1}$ are all sequentially adjacent values (e.g., RTs on Trials 1 and 2, 2 and 3, etc.) and $n$ is the total number of trials. I have calculated MSSD for two-choice RT data based on 31 children each given 30 trials; the average von Neumann ratio was 1.92 (a value consistent with randomness of the RT data). The between-subjects correlation ($r$) of $\delta$ (i.e., MSSD$^{1/2}$) with RTSD was +.69 (Spearman's rank correlation +.85), suggests that MSSD and RTSD are probably not redundant but that they tap somewhat different components of intraindividual variability. Whether the correlation of MSSD with other psychological variables differs significantly from their correlation with RTSD has not yet been determined. In making such a determination, one should also look at Spearman's rank correlation ($r_s$) in addition to the Pearson $r$, as $r_s$ is invariant for all monotonic transformations of the scales of the correlated variables.

Randomness need not be viewed as an all-or-none condition, but can be treated as a continuous variable, in which case it is perhaps better called *behavioral entropy*. However, it is important first to establish the existence of reliable individual differences in several nonredundant measure(s) of this entropy in case it is a multidimensional phenomenon of which RTSD and the von Neumann ratio reflect only certain components. One would not bother with such analysis except for the fact that intraindividual variability in RT, independently of RTm, has interesting correlations with other behavioral and physiological variables.

### Skewness of an Individual's RT Distribution

As previously explained, the distribution of an individual's RTs is typically skewed to the right, even after outliers have been eliminated by the "trimming" rules previously described. Some, but not all, of this skewness is a part of the total variance of the RT distribution, but as this skew component itself shows reliable individual differences independently of RTSD, it should be measured in its own right. A marked skewness of the RT distribution was first discovered to be one of the most distinguishing features between the RT performance of mentally retarded persons and those of average or higher levels of intelligence, as shown in Figure 4.2. It has since been found that skewness is negatively correlated with IQ throughout the whole distribution of general intelligence.

A simple but crude measure of skewness is Sk $= 3$(Mean $-$ Median)/SD. A more precise and the preferred measure, included in most computerized statistical packages, is R. A. Fisher's $g_1$ statistic: $g_1 = \Sigma(X - \overline{X})^3/N(\text{SD})^3$.

*Kurtosis*, or the degree of peakedness (or flatness) of a distribution is measured by Fisher's $g_2$ statistic: $g_2 = [\Sigma(X - \overline{X})^4/N(\text{SD})^4] - 3$. (This, too, is provided in most statistical packages.) Although Fisher's measures of skew and kurtosis are both used in evaluating models in the experimental psychology of RT (e.g., Luce, 1986, p. 38), I have not come across any mention of kurtosis in the differential psychology of RT. To begin with, it would be worth knowing if there are reliable individual differences in kurtosis ($g_2$) of the RT distribution and, if so, whether $g_2$ differs reliably between RT paradigms and is related to other psychological variables, such as IQ, independently of RTm, RTSD, MSSD, and $g_1$.

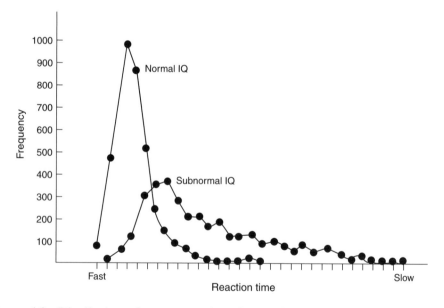

Figure 4.2: Distribution of mean reaction times of individuals with normal and subnormal IQs. (From Baumeister, 1998. Reprinted with permission from Elsevier.)

### The Slope of RTm (or RTmd) across Increasing Levels of RS Complexity

RT is often measured at two or more levels of complexity of the RS in the presented task. The simple difference between SRT and two-choice RT can be regarded as a slope measure because the straight line connecting their means slopes upward (e.g., Figure 2.3). In CRT paradigms with more than two choices or levels of complexity such as the Hick and Sternberg paradigms, in which there is a linear relationship between a subject's RTm and the various levels of the RS (e.g., Figures 2.5, 2.6, 2.9, and 2.11), it is of theoretical interest to measure the *slope* of the linear relationship. Theoretically, the slope is considered to be a measure of the rate (or speed) of information processing.

Slope is measured as the *regression coefficient*, $b$, in the equation for the linear regression of an individual's RTm (or RTmd) on a measure of variation in RS complexity, such as *number of digits* (or words) in the Sternberg paradigm or the *number of bits* in the Hick paradigm.[2]

Unfortunately, $b$ measured on individuals is probably the least reliable of the variables that have been used in chronometric research. Theoretically, as a measure of speed of information processing, $b$ has been *hypothesized* to be correlated with IQ, or psychometric $g$, even more so than are RTm, RTSD, or the regression constant $a$. But the correlation between IQ and $b$ is typically quite small and usually nonsignificant. It is therefore not a good measure of individual differences. This is not because the theory of its relation to IQ is wrong, but because $b$ has inherently low-reliability. This theoretical prediction, however, is clearly borne out by comparing the average regression lines based on groups of subjects that differ in mean IQ, because the errors of measurement in $b$ measured in individuals are averaged out in the mean of aggregated data (Jensen, 1998a). Because the RT intercept, $a$, reflects the mean of the individual's RT trials over all the levels of RS complexity, it carries most of the individual differences variance and is therefore the most reliable measure derived from the RT disribution. Hence, $a$ always has a higher correlation with any external variables than does the less reliable $b$. Another cause of the poor correlations of $b$ with IQ or other variables is its high *negative* correlation with $a$ due entirely to their correlated errors of measurement. (In the Hick paradigm, e.g., errors of measurement in $a$ and in $b$ are correlated $-.80$.) Because $a$ and $b$ share the same errors of measurement, any measurement error that increases $a$ necessarily decreases $b$ and vice versa. This means that each regression parameter, $a$ or $b$, necessarily acts as a hidden suppressor variable on the other parameter's correlation with any external variable (e.g., IQ). Therefore, if these regression parameters are used to discover relationships of RT with external variables, they should be used in combination to obtain their multiple correlation, $R$, with the external variable. The value of $R$ in this case is always larger than the Pearson $r$ based on either $a$ or $b$ alone.

### The Measurement of Errors

Error responses in tasks involving CRT consist of the subject's making the wrong response for the given RS. The presence of a low error rate, say, 2–5 percent, depending on task complexity and $S–R$ compatibility, indicates that the subject is responding at a near optimal speed in accordance with E's standard instructions to respond both quickly and accurately. Except for SRT, a total absence of errors in a large number of trials of some complex form of CRT suggests that the subject has overly sacrificed speed for accuracy. An unusually high error rate may indicate the opposite. In individual difference research on RT, it is important to keep the *between-subjects variance* in error rates at a minimum, as the variable of primary interest is RT. This is seldom a problem, and average error rates generally vary little if at all even between groups that show extreme differences in CRT, such as retarded and normal adults, or younger and older children. In fact, we have found that, except when a great many trials are administered, the typically low error rates for various CRT tasks have low or erratic reliability. Error rates have shown near zero correlations with psychometric *g* in a college population (Kranzler, 1991). When the overall error rate is above 5 percent, either for an individual or for a subject sample, one must question whether the task is too difficult to be a good measure of CRT for those subjects.

Errors should always be counted. But because error responses may affect the RTs for which they occur, more reliable measures of RTm or other parameters of the RT distribution would be obtained by basing these measures only on error-free trials. So that the RT measures are based on the same number of error-free trials for every subject, the procedure we have often used is to *recycle* the particular RSs on which errors occurred, presenting them at the end of the sequence of previously programmed trials. This procedure (which also counts the number of recycled trials) is programmed as a part of the computerized administration of the RT test.

A psychometrically objectionable practice is to create an overall composite "performance score" based on some formula for combining the individual's RTm with the total number of errors. RT and errors are distinct variables. If there should be any reason to adjust individual differences in RTm or other RT parameters for error rates, this is best accomplished by statistical regression or partial correlation.

### *SRT as a Suppressor Variable in Correlations of CRT with External Variables*

The SRT paradigm evokes minimal uncertainty and central decision processes, namely the subject's uncertainty as to the precise moment of onset of the RS. A prepared and practiced subject has zero uncertainty and zero choice concerning the required response. Therefore a larger fraction of the subject's SRT consists of the time needed for the sensory transduction and neural transmission of the RS to the brain and thence for the neural transmission to the muscles, plus muscle lag, than exists for any type of CRT. CRT paradigms involve all these sensorimotor components and the temporal uncertainty of the RS onset that compose the SRT *plus* the time taken for other central decision processes demanded by the CRT paradigm. As CRT tasks increase in information processing demands, there is a regular decrease in the correlation between CRT and SRT. SRT and 2-choice RT have about 80 percent of their true-score variance in common; SRT and 8-choice RT have only about 50 percent in common (Jensen, 1987a, p. 139).

If a researcher's interest is primarily in the correlation of individual differences in the central information processes tapped by CRT with other measures of cognitive activity, then this correlation is attenuated by the sensorimotor component of CRT. In examining the correlations of IQ with SRT and CRTs for different levels of complexity (8-choice Hick and Odd-Man-Out paradigms), it was found that SRT acts as a suppressor variable in the correlation, so that removing the SRT component from CRT raises the correlation of CRT with IQ (Jensen & Reed, 1990). Three methods for ridding the correlations between CRT and an external variable of the attenuating and suppressing effect of the sensorimotor components of CRT were examined by Jensen and Reed, as follows:

(1) Subtract every subject's SRTm from CRTm. The disadvantage of this method is that difference scores have much lower reliability than do the separate elements, especially when the elements are highly correlated (as is the case with SRT and CRT).
(2) Use partial correlation, partialing SRT out of the correlation between CRT and the external variable.
(3) Include both SRT and CRT in a multiple correlation ($R$) with the external variable. This has the advantage of picking up the information processing component of the SRT, while suppressing its sensorimotor component. In other words, the method of using a multiple $R$ suppresses the sensorimotor components that the predictors (i.e., SRT and CRT) have in common with each other but which neither has in common with the external variable (e.g., IQ), while the proportion of variance in the information processing components that SRT and CRT both have in common with the external variable is given by $R^2$.

All three methods are effective, but they differ in their effectiveness depending on the reliability of the SRT and CRT measurements. Generally, the higher the reliability of SRT, the more similar are the results of the three methods. Based on a detailed statistical and empirical consideration of the use of SRT as a suppressor variable in correlational studies of CRT, Jensen and Reed (1990) concluded "We recommend that a measure of simple RT be obtained in all studies of the relationship between complex chronometric variables and psychometric *g* or other abilities, and that the relationship between the chronometric and psychometric variables be examined not only by means of their zero-order correlation, but also when variance in SRT is directly or statistically controlled to rid the correlation of possible attenuation by peripheral, noncognitive factors" (p. 388).

In all chronometric research it would be most advantageous to obtain highly reliable measures of SRT and the corresponding movement time (SMT) under completely standardized conditions for every subject, regardless of whatever other chronometric paradigms are used. This would serve not only the statistical purpose described above, but the descriptive statistics of the subject sample's SRT and MT distributions help in describing the sample and locating it in the dimensional space of RT parameters that would permit direct comparisons with samples in other chronometric studies. The importance of the universal standardization of every aspect of chronometric methods, apparatus, and procedures can hardly be emphasized strongly enough.

### MT as a Separate Variable

Although many studies have measured RT without controlling MT, thereby measuring some unanalyzable amalgam of both, there are good reasons for measuring RT and MT separately. Having the subject release a HK and pressing a RK helps to divide the subject's total response into two separately timed components, RT and MT. These components have markedly different characteristics, which have been documented elsewhere, with reference to the Hick paradigm (Jensen, 1987). Whereas RT increases regularly with an increase in complexity or information processing demands of the RS, MT shows little, if any, systematic change across increasing levels of task complexity. For the same tasks, RT is invariably longer than MT (although subjects' subjectively perceive just the opposite). The within-subjects correlation between RT and MT over trials is virtually zero. The between-subjects disattenuated correlation between RTm and MTm averages .55 in young adults and increases in elderly adults. RT is generally more correlated with psychometric *g* than is MT, and the correlations of RT and MT with IQ vary independently with age, being higher in young children and elderly adults and much lower in young adults. While RTSD is substantially correlated with *g*, MTSD is not. What is most striking is that the correlation of RT with IQ *increases* with greater task complexity, whereas the correlation of MT with IQ *decreases* (Jensen, 1987a, pp. 164 – 165).

When seven measures of RTs and MTs obtained from various paradigms were factor analyzed along with an IQ battery consisting of 11 psychometric tests, the orthogonalized hierarchical factor analysis clearly separated all the RTs and MTs, which had their salient loadings on two orthogonal factors (Carroll, 1991a,b). The average factor loading of RTm with the general factor in the whole battery of chronometric and psychometric tests was $-.52$; the average loading for MTm was $-.13$. (All of the psychometric tests had salient loadings on the general factor, and the verbal and nonverbal psychometric tests had salient loadings separately on two lower-order group factors.) It appears that CRT reflects mostly information processing whereas MT reflects mostly sensorimotor ability. All these differences noted between RT and MT argues for the importance of measuring both RT and MT and treating them statistically as distinct variables.

### Reliability of Chronometric Measures

Reliability refers to the consistency of individual differences in a particular task measurement (RT, RTSD, MT, etc.). The split-half reliability coefficients based on odd–even trials are high (around .90 with as few as 30 trials) for both RTm and MTm, and test–retest reliability for 15 trials measured on each of two days is about .80 for RTm and MTm. The split-half reliability is almost perfectly a function of the number (*n*) of trials. It conforms better to the well-known Spearman–Brown formula[3] for predicting the reliability of a test of a given length, *n*, than do most psychometric tests. The Spearman–Brown formula is therefore useful for estimating the number of RT trials needed for any desired reliability of RTm or Mtm.

Because the reliability of a test cannot be higher than the square root of its correlation with any other variable, we may infer a lower-bound estimate of reliability from the correlation between different measurements. Hence an indirect indicator of the reliability of RT measurements, though not a measure of reliability *per se*, is the correlation between RTs measured for different set sizes of the Hick paradigm. In 11 samples totalling 941 subjects, the average Pearson correlation between RT for every set size (*n* = 1, 2, 4, 8) and every other set size was .79, estimating an average lower-bound reliability of $\sqrt{0.79} = 0.89$ (Jensen, 1987b, p. 141, Table 17). The correlations between RTs obtained from different RT tasks is still another indicator of the lower-bound reliability of the measurements reliability. In a study of seven different RT paradigms given to 40 college students, the average correlation between the RTs obtained on the various paradigms was 0.59, estimating an average lower-bound reliability of $\sqrt{0.59} = 0.77$ (Hale & Jansen, 1994).

When repeated measures of RT are obtained on different days or even on the same day at different times separated by a few hours, the test–retest reliability is generally slightly lower than odd–even reliability obtained within a single test session. Reliability is higher in more practiced subjects. In any case, with sufficient practice trials and a large number of test trials, obtained if necessary over two or more sessions, virtually any degree of reliability can be achieved. Of course, reliability need not be greater than is required by the specific purpose of the particular study. Comparisons of the same person's RT performance at different points in time, or on different RT paradigms, calls for higher reliability than comparisons between individuals, which requires higher reliability than comparisons between groups of individuals.

## Notes

1. The reliability of a difference between any two scores, say $A$ and $B$, is $r_{(A-B)(A-B)} = (r_{AA} + r_{BB} - 2r_{AB})/(2 - 2r_{AB})$.
2. The predicted $\hat{Y}$ (e.g., RTm) as a linear function of X (e.g., the quantized level of RS complexity) is: $\hat{Y} = a + bX$, where $a$ is the *intercept* (or regression constant) and $b$ the *slope* (or regression coefficient).
3. Given the known reliability $r_{xx}$ of a test consisting of $n$ items (or trials, in the case of an RT test), the predicted reliability, $r_p$, of a test with $k$ times as many similar items is $r_p = kr_{xx}/[1 + (k-1)r_{xx}]$.

# Chapter 5

# Chronometry of Mental Development

Chronometric methods can be especially valuable, and are often essential, for the study of developmental changes in cognitive aspects of behavior from infancy to maturity. Its two major advantages are lacking in all other types of psychological assessment. First, both the *independent* variable (age) and the *dependent* variable (behavior) are direct measures of time, based on one and the same ratio scale, whether expressed in units of milliseconds, months, or years. Second, many of the elementary cognitive tasks (ECTs) that can be used to measure response times can be identical across a much wider age range than is possible for other types of tests based on item performance scored as right/wrong or pass/fail. In such tests, floor and ceiling effects are narrowly separated; as a result, any given set of the same test items cannot be used to measure individual differences outside an age range of more than a few months. Older subjects must be tested on different and more difficult items than younger subjects, and it is always problematic whether the different items across a wider age range are scaled in equal intervals of ability or difficulty. Many chronometric tests, however, can yield highly reliable measurements across an age range from early childhood (EC) throughout the human life span, and across a range of ability from the mentally handicapped to the exceptionally gifted. Moreover, aside from the response times, response errors are more causally interpretable within a theoretical framework than are the errors made in most conventional tests, because the subject's possession of the specific knowledge components required by the task can be easily ensured for most chronometric tasks.

In view of its advantages, it is surprising how little use has been made of chronometry in research on children's mental development, which has depended mostly on observational, clinical, and psychometric methods. However, there have been a few notable developmental studies in which time measurements were central. My aim here is to point out some of the commonalities and regularities in the measures found in a few of these studies based on well-known chronometric paradigms, but without detailing the specific apparatus and procedures.

One might ask why anyone should be interested in the application of chronometry to developmental research. The answer is that chronometric measures are known to be correlated with central variables and concepts of interest in child development, such as mental age, IQ, spurts and lags in mental growth rates, and the sequential emergence of special abilities differentiated from psychometric *g*.

Readers who would delve into this literature should be aware of the lack of *methodological standardization* across the various studies even when based on the same chronometric paradigm. This is one of the misfortunes in this field at this point in history. Although many studies are consistent in the general features of their findings, they are not directly comparable in the absolute values of the obtained measurements.

I have had the experience of using three slightly different apparatuses for measuring reaction times (RTs) and movement times (MTs) in the Hick paradigm, and have been surprised by the differences they yielded in the absolute values of RTs and MTs, although the typical results showing Hick's Law were obtained with all three apparatuses. In visiting various psychological laboratories in the United States, Canada, Asia, and Europe, I have found even greater variations in the apparatuses used for studies of any given well-known RT paradigm. Yet there have been exceedingly few parametric studies of the constant differences in the obtained measurements. Therefore, when different age groups have been tested on the same paradigm (but on different apparatuses) in various laboratories, the results of the different groups cannot be meaningfully plotted on the same graph. It is impossible, then, to distinguish differences between subject samples from differences in apparatus and procedures. Consequently, the comparability and possible cumulation of first-order data from different laboratories is seldom possible. This is true even though all of the measurements obtained in each laboratory are perfectly accurate. And generally, the essential phenomena demonstrated by a given paradigm are manifested regardless of variations in the apparatus and procedures. But the differences in apparatus (and often in the testing procedures) create marked differences in the absolute measurements. Under such a handicap of nonstandardized methods of measurement, the physical sciences could never have advanced to their present level. All that present-day psychologists can report from the studies performed in different laboratories is scale-free correlation coefficients and the general shapes of growth curves obtained from either longitudinal data (i.e., same individuals measured at different ages) or cross-sectional data (i.e., different but comparable groups differing in age).

A feature of the chronometric literature on mental growth that is not discussed in detail here is the attempt to explain causally and theoretically the empirical finding of increasing speed. What often passes for an explanation of a behavioral phenomenon is better considered simply as a more detailed and elaborated description of the behavior rather than as a causal explanation. Slower RT in younger children, for example, is described as the result of their greater distractibility, poorer attention span, tendency to fidget after a few test trials, and the like. But the signs of such behavior may be expressions of the same causal system of which slower RT is another manifestation. Behavioral changes accompanying mental development are expressed in many forms of nonspeeded behavior as well, such as performance in Piaget's tests of conservation, the ability to copy certain simple geometric figures, or to recall more than, say, three digits in the digit span subtest of the Stanford-Binet. RT is correlated with each of these developmental tests, even when age is statistically controlled. So there is a common factor in the age-correlated tasks that are not explainable in terms of any particular set of purely behavioral characteristics.

Regarding age differences in cognitive abilities between infancy and adulthood as well as individual differences in general, experimental psychology historically has perpetuated a tabula rasa bias of attributing them exclusively to variables originating externally to the subjects. Age trends in cognitive capabilities thus tend to be viewed as wholly experiential, a product of learning, rather than as the developmental expression of genetically programmed biological processes. Much of the evidence presented later on in this book contradicts the peripheral or behavioristic view of cognitive development as mainly controlled by external events. This view is contradicted, in general, by what might be called the relative *experiential neutrality* of the many chronometric measurements that display large age differences as well as individual differences independent of age.

## Response Time Differences Across the Human Life Span

Although this chapter deals with the time taken for various cognitive activities in the developmental segment of the life span, it may be useful first to view them in relation to the trend of response times across the whole life span, from the earliest age that RT has been reliably measured to senescence (S). Even though no single study covers the whole life span using a single chronometric paradigm with a standardized apparatus, by examining data from many studies in which this ideal condition prevails for different segments of the life span, we can infer a generalized composite trend of relative response times across the full span (see Cerella & Hale, 1994; Welford, 1980c; Wickens, 1974). As shown in Figure 5.1, the generalized RT curve reaches its lowest point in the mid-20s, but very gradually increases in the late 20s; the 30s are already noticeably slower, and the rising curve increasingly becomes positively accelerated while approaching later maturity and senescence (which is the subject of Chapter 6). Figure 5.1 is the averaged or generalized trend line around which there is naturally much individual variation. Yet, it is safe to say that virtually no long-lived persons fully escape the toll that increasing age takes on mental speed. Figure 5.2 shows a less idealized and perhaps somewhat more idiosyncratic graph based on cross-sectional RT data from 10 age groups in 11 different studies. Because diverse RT paradigms were used in the various studies, the age differences in RT have to be expressed on a relative scale called the *RT ratio*. For any given paradigm, this is the ratio of a given age group's mean RT to the mean RT of the group with the shortest mean RT (in this case, ages 20–29) and multiplied by 100. Thus, the mean RT ratio at the mid-point of the ages

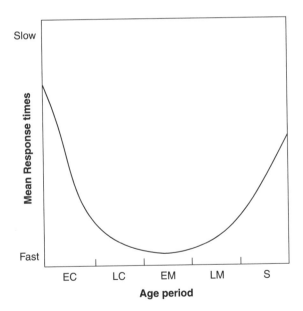

Figure 5.1: Generalized trend of response times across the life span in five age periods: infancy and EC, LC and adolescence, early maturity (EM), late maturity (LM) and senescence S.

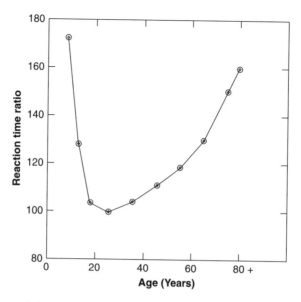

Figure 5.2: Age trend from age 6 to 80+ based on 11 cross-sectional studies of 10 age groups and using various RT paradigms expressed as RT ratios for each age group, based on data assembled by Welford (1980c, Table 9.2, p. 330). The RT ratio is the mean RT in a given age group divided by the mean RT in the age group with the shortest mean RT (i.e., aged 20–29) and expressed as a percentage. The data points are plotted at the mid-points of the age intervals in years 6–9, 10–14, 15–19, 20–29, 30–39, 40–49, 50–59, 60–69, 70–78, and 80+.

20–29 interval is 100 percent, while the mean RT ratio at the mid-point of the ages 6–9 interval is 172 percent.

Most of the RT research in recent years has been done using undergraduate college students in the age range of 18–22 years. This population, besides its youth, has IQs almost entirely in the upper half of the population distribution. It obviously represents a very restricted range of RT variation compared with that in the total population. Therefore, the range restriction in college samples on both age and IQ (which also is correlated with RT) also restricts the variance in RT, a condition that should always be taken into account in interpreting any chronometric studies based on college students.

Every study of timed performance indicates that from infancy to early and later childhood (LC) there is a conspicuous decrease in RTs and a corresponding increase in speed, even in many activities that are not considered traditional chronometric paradigms, such as the speed of reaching for an object, counting things, and repeating words (Case, 1985, Chapters 16 and 17). Figure 5.3, for example, shows the mean times taken for cross-sectional samples of Canadian school children, aged 6–12, to count the number of green dots on a card, with yellow dots as distractors. A total of 50 green dots was divided among eight separate test cards. Subjects were instructed to touch each green dot with their index finger while counting the dots as quickly and accurately as they could. Sufficient practice

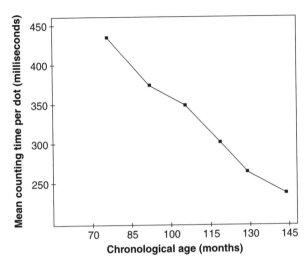

Figure 5.3: Mean counting time (time/dot) for children aged 6–12, in Ontario, Canada. (Adapted from Case (1985), Figure 17.1.)

trials were given to ensure that everyone could meet the task requirements. The test did not require counting beyond 10, which even the youngest children were able to do at the outset. Counting time was measured as the average time per dot, based on the total number of correctly counted dots. Between ages 6 and 12, the slope of counting time as a function of age is nearly linear.

On some of these tasks that call upon some prior knowledge, of course, age differences, especially in EC, reflect an unanalyzed combination of prior practice and maturation. To minimize the influence of prior knowledge on performance, one needs to look at age differences in some of the classical RT paradigms, which are less experience based and therefore mainly reflect maturational changes, whatever their presumed biological underpinnings. To analyze the relative effects of experiential and maturational factors in response times, chronometric experiments can be specially devised to introduce varying demands on prior knowledge (e.g., using verbal and numerical stimuli) or the degree of familiarity with the specific stimulus and response requirements of the information-processing task. A few selected examples can illustrate age effects on some classic RT paradigms in which the very same paradigm and apparatus were used in every age group. The details of the particular paradigm, apparatus, and procedures are given in the cited studies.

### The Hick Paradigm

Data from studies of the Hick paradigm in my laboratory, summarized elsewhere (Jensen, 1982, 1987, 1989a) show RT differences across a fairly wide age range, the main virtue of the data being that they were obtained with the same apparatus (the "Jensen box") with the same procedures, and usually with the same tester, as for the cross-sectional results shown in Figure 5.4. The 0- and 3-bit conditions correspond to 1 and 8 light/buttons, respectively,

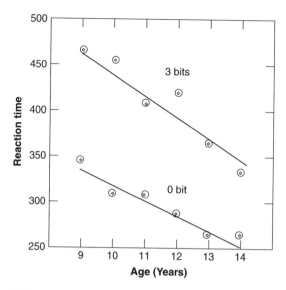

Figure 5.4. Mean RT (in ms) in responding to a response stimulus representing 0 and 3 bits of information (i.e., 1 light/button or SRT, and 8 light/buttons, respectively, on the Hick paradigm with the Jensen-box apparatus) for school children averaging 9–14 years of age.

with the same subjects (from the same school) given both conditions. Over the age range of 5 years during late childhood, the RT data do not depart significantly from a linear trend. The 3-bit task has a steeper slope on age than the 0-bit task. Similar results were found in another study of the Hick paradigm showing RT differences between seven age groups from ages 5 to 12 years (Fairweather & Hutt, 1978). However, in this study the absolute differences in overall mean RTs were considerably greater and the differences between ages 5 and 10 years appear less pronounced, probably because RT and MT were not measured separately, but consist of a combined measurement nominally called RT, obtained by having the subjects press one of 10 keys over which the fingers of each hand were placed. The 5-year-olds' 4-choice nominal RT (actually some combination of RT+MT), for example, was about 1200 ms as compared with subjects of the same age using the Jensen-box Hick apparatus, with about 700 ms for 4-choice RT and about 640 ms for MT. (This is another example of how differences in apparatus and procedure can rather drastically affect the absolute values of nominally the same measurements of processing time.)

Figure 5.5 shows Hick data, using the same Jensen-box, for different age groups spanning 15 years, from preschool and kindergarten to college, with mean RT plotted as a function of the mean age of each group. Only 0, 1, and 2 bits are shown, because many of the preschoolers had trouble with the 3-bit condition and made too many errors for their RT to be validly compared with that of the older groups. To reduce the ambiguity of RT comparisons, only the 22 out of 30 children in the youngest group whose error rates for 0, 1, and 2 bits were fairly comparable to that of the older groups were included in these data. Even so, it is evident that an enormous difference in information-processing time occurs between

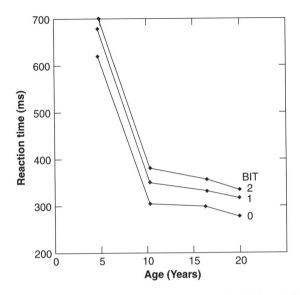

Figure 5.5: Mean RT on Hick paradigm (with Jensen-box) for 0, 1, and 2 bits (i.e., 1, 2, or 4 light/buttons) in age groups from 5 to 20 years.

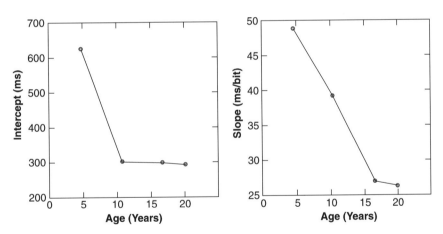

Figure 5.6: Mean intercept and mean slope of RT of the Hick function plotted as a function of age, for the data shown in Figure 5.5.

5 and 10 years of age; the RT differences beyond age 10 are real, but relatively small, with RT decreasing, on average, at the rate of about 3.2 ms/year, as compared with a rate of about 65 ms decrease/year between ages 5 and 10 years. The average processing speed of the 5-year-olds is 25 bits/s; for the 20-year-olds it is 36 bits/s, or 1.44 times faster than that of the 5-year-olds. Figure 5.6 shows the intercepts and slopes of the Hick function in these four age groups. The great difference between the shapes of these functions at about age 10

indicates that the developmental trends of these two aspects of RT are on different age trajectories. This is also evident that different processes are involved in the intercept and the slope of the Hick function. The *intercept* is believed to be the time taken by processes such as stimulus apprehension, perceptual encoding, response preparation, and muscle lag, whereas the *slope* is the rate for processing the information load *per se*, i.e., the reduction of uncertainty as to the correct response when the reaction stimulus (RS) occurs, expressed as the number of milliseconds per bit of information. On the Hick paradigm, age groups above 10 years show small intercept (stimulus encoding) differences relative to slope differences (see Figure 5.6). If the demands on stimulus encoding were made considerably greater than they are in the Hick paradigm, the reverse could be true, i.e., the age differences in slope could be made negligible by comparison with the intercept differences.

### Sternberg Paradigm

The age effect of a greater demand on stimulus encoding has been demonstrated in two experiments by Maisto and Baumeister (1975), using the S. Sternberg memory-scan paradigm. One of these experiments is described here to illustrate the effect of an experimental manipulation of the test stimuli on intercept and slope as a function of age. The stimuli were eight line drawings picturing highly familiar objects (car, airplane, TV, fish, dog, house, sailboat, lamp). Half of the stimulus items constituted the fixed *positive set*, which was thoroughly memorized by all subjects. Set sizes of either 2, 3, and 4 pictures were used. On a random half of the trials, a single-probe stimulus was a member of the positive set. The subject responded to it verbally "Yes" or "No" (with RT measured by an electronic voice key) as quickly and accurately as possible as to whether the probe was a member of the memorized positive set. The subject's RT (yes or no) was measured by a voice key. The probe stimuli were two types of pictures: *intact* or *degraded*, of which examples are shown in Figure 5.7. The subjects' mean ages in years were Preschool (4.9 years), 3rd Grade (8.8 years), and 5th Grade (10.5 years). The RT results in Figure 5.7 indicate large and significant *intercept* differences (on the ordinate) between grades and between stimulus type (*intact/degraded*), and there are significant differences between set sizes. The differences in slopes, associated either with Grade level or with stimulus type, are statistically nonsignificant and of negligible magnitude. A perfectly parallel experiment by Maisto and Baumeister (1975) using digits instead of pictures as stimuli showed quite similar results, except that the degraded stimulus condition (a grid placed over the digits) had a significantly lesser intercept effect for 5th Graders than for the two younger groups. The differences in slopes were negligible for Grades and for stimulus types. Hence, in the Sternberg paradigm using these stimuli, the time taken by stimulus encoding and response preparation, reflected by the RT intercept, is shown to be on a much steeper age trajectory during the EC years than is the time taken by memory scanning, reflected by the slope of RT on set size.

Degrading a stimulus, however, increases the perceptual time and hence also the encoding time. But the perceptual process *per se* has exceedingly little effect on the speed of memory scanning, i.e., the search for the probe stimulus in the previously memorized positive set. Still another pair of experiments based on the Sternberg paradigm asked whether the slope of RT on set size could be importantly affected by an experimental manipulation of a variable having more purely cognitive potency, namely, the

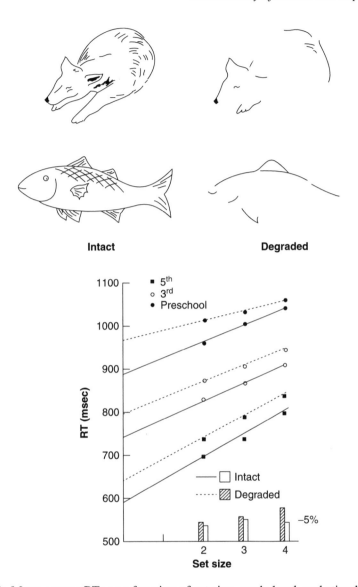

Figure 5.7: Mean correct RTs as a function of set size, grade level, and stimulus quality. (Examples of *intact* and *degraded* stimulus quality are shown above.) Lower bars are error rates. (Reprinted from Maisto and Baumeister (1975), with permission of Academic Press Inc.)

meaningfulness of the stimuli (Baumeister & Maisto, 1977). One stimulus type consisted of eight pictures (simple line drawings) of *familiar objects*; the other stimulus type consisted of line drawings of rather simple but distinctive *random figures*. The testing procedure was the same as in the previous experiments, and the age groups again were

Preschoolers (5.5 years), 3rd Graders (9.1 years), and 5th Graders (11.6 years). Subjects answered vocally "Yes" or "No" as to whether or not the probe stimulus was a member of the previously memorized positive set, while the RT was registered by a voice key. The main result was that the random figures showed much greater slopes of RT across set size in all three age groups than did the familiar figures, and in each age comparison the younger group consistently shows a steeper slope than the older group. In another experiment, when subjects were instructed to verbally mediate their memorizing of the random shapes by giving them names of familiar objects that they could be imagined to resemble, the slope of RT on set size was markedly lessened as compared with a noninstructed group. The slope, or rate of short-term memory scanning, is determined much more by centrally mediated cognitive processes in memory search than by perceptual intake processes.

### Inspection Time: The Maturation of Perceptual Intake Speed

Inspection time (IT) measures the speed of perceptual intake. The time interval represented by IT is, on average, much shorter than that for measures of simple reaction time (SRT). (The IT paradigm is described on pp. 33–35.) The advantage of IT, as opposed to RT, is that IT eliminates the entire motor component of response time. The subject only has to make a simple visual discrimination (e.g., which of two lines displayed tachistoscopically for a given time lines is the longer?) and may take as much time as needed to respond, the response time itself being wholly irrelevant. Hence, IT affords a sensitive means for studying the developmental trend in speed of perceptual intake independently of the whole efferent aspect of RT.

The first studies of the development of IT during childhood (Nettelbeck & Wilson, 1985; Wilson & Nettelbeck, 1986; Nettelbeck, 1987) showed an age trend much like that for RT, as seen in Figure 5.8. It was also found in longitudinal studies that practice on IT

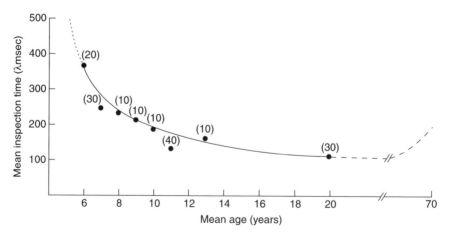

Figure 5.8: Mean IT as a function of age. (Reprinted from Nettelbeck (1987), with permission of Abelx.)

up to asymptotic levels of performance resulted in a much smaller decrease in mean IT than the mean decrease associated with a 2-years increase in age. That is, the major improvement in IT appeared to be more due to maturation than to practice effects. Certainly few, if any, of the subjects practiced, anything even remotely resembling the IT paradigm during the intervening years. No surprises here; the picture for IT is much as one would expect from all the developmental studies of RT.

But then a more recent study, clever and seemingly impeccable in design, has now added a new element of mystery to the developmental aspect of IT (Anderson, Reid, & Nelson, 2001). As they used an unconventional IT task, deemed more suitable for sustaining children's interest, further studies will have to determine the extent to which their novel IT stimuli might account for their rather surprising results. Using a combined cross-sectional and longitudinal design, this research team tested four different age groups (aged 6, 7, 8, and 9 years) on the IT paradigm on each of three successive years (years 1, 2, 3). In the first year's testing, a retest was administered after a 5-min rest interval, to determine test–retest reliability and the effect of practice. This short-term practice effect was very small and statistically nonsignificant. However, the more important finding is shown in Figure 5.9, which compares age-group differences with the retest of each group at 1-year intervals. The overall differences between age groups in IT are significant, as are the overall differences at 1-year intervals within each of the age groups. But here is the surprise: the effect on IT of retesting after 1 year is much greater than the effect of an age difference of 3 years. For example, when the 6-year olds are tested for the third time (at age 9), they have a much shorter IT than do the 9-year olds tested for the first time at age 9. The truly developmental effect (i.e., the age-group

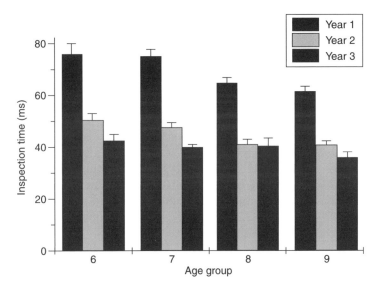

Figure 5.9: Mean ITs as a function of age group and year of testing. Error bars = one standard error of the mean. (Reprinted from Anderson, Reid, and Nelson (2001), with permission of Elsevier Science Inc.)

differences at year 1) is small compared with the mysterious "practice effect" (i.e., the retest differences between years 1 and 2). But what is this mysterious "practice effect"? Is it a kind of *reminiscence* effect, which is an improvement in performance following a relatively long rest interval? The study's authors refer to it as a "conundrum" and ask ". . . what is learned such that immediate practice effects are minimal but yearlong carry-over effects are huge?" They hypothesize that (a) the improved IT performance may have resulted from the children's becoming more familiar and relaxed in the testing room, where they also took many other tests; and (b) the initial testing instigates a process of "slow changes and reorganization in underlying knowledge and strategic behavior." Yet, there are consistent age differences at the first time of testing that need to be explained, whatever may result from the subsequent long-term "practice effect." The authors concluded,

> What is true is that there is a great deal more room than we believed before this study for the possibility that what *changes* with age are generalizable strategic processes. The complement of this conclusion should also be borne in mind. We must be cautious in interpreting differences in speed measures such as IT as indicating differences in the underlying theoretical construct 'speed of processing.' Were we to do so in this case we would find ourselves in the uncomfortable position of conceding that task-specific prior experience can reduce speed of processing. This would seriously weaken the utility of what has proved to be an interesting and fruitful theoretical construct. (p. 485)

## Age Differences in the Effects of Procedural Conditions

Except for the overall relative slowness of RT for younger children, it is risky to generalize the details of any given paradigm or experimental procedure to others. Even rather subtle variations in these factors have effects on RT that are generally of greater magnitude in younger than in older children. This argues especially strongly for strict standardization of chronometric procedures in studies of preadolescent subjects if meaningful comparisons are to be made across different studies of a particular phenomenon.

Complex factorial experiments on the effects of varying several experimental conditions have been performed by Rogers Elliott (1970, 1972), using several age groups from 5 to 13 years and young adults (college students). In all studies, the RT task was SRT to an auditory stimulus. The five independent variables were: (1) *age* (years), (2) *preparatory interval* (PI = 1, 2, 4, 8, or 16 s); (3) *PI constant or varied* within a test session; *incentive* (monetary rewards for faster SRT); (4) *incentive-shift* (various combinations of shifting from a high to a low reward, e.g., HL, LH); and (5) *amount of practice* (1–10 sessions). Variation in all four of these conditions produced significant main effects, often surprisingly large. Within age groups most of the first-order interactions among the experimental variables were also significant. The results of such a multifactorial study with all its interactions are quite complex; they are most easily summarized graphically, as shown in Figure 5.10.

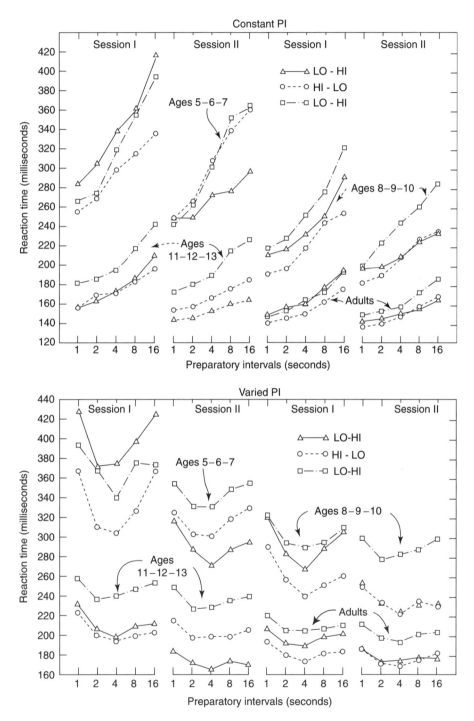

Figure 5.10: Auditory SRT as a function of five independent variables explained in the text. (Reprinted from Elliott (1970, p. 94), with permission of Academic Press Inc.)

## Interaction of Response Errors and RT as a Function of Age

The central question in developmental chronometric research is discovering the cause of the marked increase in speed of information processing between EC and adulthood.

Two classes of hypotheses have been considered: (1) age-related limitations of processing speed imposed by structural developmental changes in the brain, and (2) age-related increases in task-relevant knowledge, attention, and an enlarging repertoire of increasingly efficient control processes associated with strategies for monitoring one's performance. Both classes of causation are most likely involved in any given task, but their relative importance would depend on the specific task requirements. Explanations in terms of purely psychological constructs remain vacuous until they can be operationalized and measured independently of the phenomenon they attempt to explain or can predict novel phenomena that are empirically testable. There always remains the problem of a regress of explanatory constructs. If an age-related increase in "attentional resources" is claimed to explain faster RT, for example, what explains the improved attentional resources? Or are improvements in attention and in RT both a result of one and the same cause, which as yet is undetermined? Without neurophysiological explanations at present, the best we can do is to refine the description of RT performance in ways that will identify collateral features of performance that afford further clues to the locus of independent developmental processes that affect age differences in processing speed.

An excellent example of such an approach, drawing on collateral features of RT performance, is a study by Brewer and Smith (1989) comparing different age groups at 2-year intervals from age 5 to 15 years and young adults (mean age 23.4 years). The task was 4-choice RT. After 400 practice trials in two sessions, subjects were given 200 trials in each of 10 sessions, for a total of 2000 test trials. Subjects responded to each of the 4-lights by pressing one of four response keys on which they placed the index finger and middle finger of each hand. Because these types of responses involve a considerable motor element in addition to the cognitive processes, the generality of the findings would be enhanced by another study using only the index finger release of a single home button, then pressing one of four response buttons with the same index finger, thereby minimizing the motor coordination element and permitting separate measurements of RT and MT.

The main hypothesis examined in this study was that RT improves with age during childhood because of developmental changes in the operation of control mechanisms that mediate speed–accuracy regulation. Younger children typically show much greater intraindividual variability in RT than older children and adults. As children mature, they get increasingly better at using feedback information on each trial to regulate their position of the speed–accuracy operating curve (see Figure 3.1) so as to zero-in on a narrower band of RTs that both reduces errors and increases speed. Attention and informative feedback regarding errors (i.e., pressing the wrong response button) was measured by having subjects indicate whenever they made an error by pressing a separate button with their left little finger. Although *error* rates (averaging about 3 percent) were not related to age, subjects' *error-detection* rates were markedly and significantly related to age. The percentage of selfdetected errors increases regularly from about 36 percent for 5-year-olds to 92 percent for 11-year-olds. The mean RTs for correct and error responses in each age group are shown in Figure 5.11; again we see the great drop in RT during EC (aged 5–9).

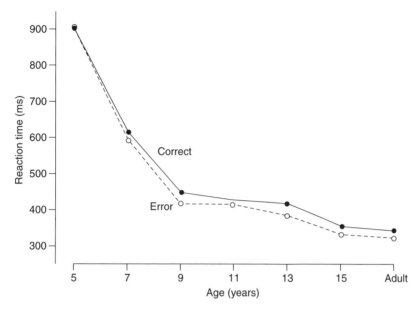

Figure 5.11: Average median correct and error RTs for each age group. (Reprinted from Brewer and Smith (1989, p. 302), with permission of the American Psychological Association.)

But the most interesting finding in this study is the nature of the RTs on the trials *preceding* and *following* an error response, which tend to average out in a subject's mean RT based on many trials. Figure 5.12 shows the mean RTs of the 10 trials preceding and the 10 trials following an error response in each age group. At every age, subjects appear to be continually testing themselves for the fastest response speed they can get away with without risking an error, and when an error occurs they immediately go back to a more cautious speed of responding. The RTs for 5- and 7-year-olds are markedly more erratic and the posterior effects of an error are much greater than in the older groups, indicating a less "fine-tuned" control of RT. Note also that it takes longer for the younger groups to recover to a faster RT after an error response; slower than average RTs occur for several trials following the error. The authors concluded as follows:

> In summary, the most dramatic age-related reductions that we observed occurred from ages 5 through 9 years. Corresponding to these changes were clearly identifiable improvements in the monitoring and regulation of speed and accuracy of performance, as well as a reduction in the lower RT limit at which accurate responding could be sustained. From age 9 years on, those control mechanisms mediating accuracy monitoring and regulation of speed of responding appeared to be operating effectively, and further RT reductions were relatively small. Analyses of error patterns pointed to improved stimulus discrimination as a likely factor underlying those additional improvements seen beyond about age 13 years. (Brewer & Smith, 1989, p. 309)

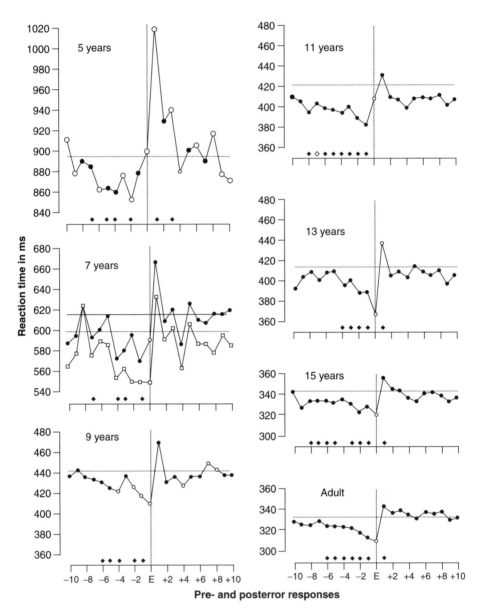

Figure 5.12: Average median correct RT for each age group on the 10 trials before and 10 trials after errors. (Diamonds indicate pre and posterior RTs that are significantly ($p < .05$) shorter and longer, respectively, than the overall median correct RTs indicated by the horizontal dotted line. (For the group of age 7 years, the squares represent the RT patterns when one outlying subject was excluded.) (Reprinted from Brewer and Smith (1989, p. 305), with permission of the American Psychological Association.)

The authors remained agnostic regarding the relative effects of intrinsic developmental structural limitations on RT and age-related changes associated with monitoring and regulation of speed and accuracy in RT performance. But the regulation of speed and accuracy is probably themselves elements of the same structural developmental trajectory as well as is the speed of information processing. So, explanation in purely psychological terms goes only as far as a more detailed description of the phenomenon and its relationship to other behavioral phenomenon, as seen in Brewers's and Smith's carefully detailed study. But the more such empirical connections that are discovered between different behavioral phenomena, the more clues neurophysiologists will have for discovering the basic causal mechanisms. The basic causes of all classes of individual differences in behavior are the operations, not of psychological constructs, but of physical structures and processes in the brain.

## An Age-Related Global Factor in Processing Speed

The most impressive programs of research on the developmental aspects of mental-processing speed are those conducted by Robert Kail and Sandra Hale. Both are leading experts in the field of experimental developmental psychology. Their research programs arrive independently at very similar and unifying conclusions. (The major publications issuing from their programmatic research are cited in Kail (1991a,b), Hale (1990), Fry and Hale (1996)). The most important point on which their findings converge is the evidence for what they term a *global* mechanism that changes with age and limits the speed of information processing across a wide variety of RT paradigms and classical developmental tasks.

Kail (1991b) expresses the rationale of his research as follows:

> Just as the central processing unit of a microcomputer can run programs in different languages that accomplish an incredible variety of tasks, the fundamental processes of cognition are almost certainly the same for all humans, despite the fact that the organization of these fundamental processes to perform more complex acts is strongly culture-bound. A working assumption in my research is that the speed with which fundamental cognitive processes can be executed may well be one of those aspects of cognition that is universal rather than culture-bound. . . . [C]ulture-bound beliefs about the virtues of performing rapidly or carefully are conceptually distinct from the maximum speed with which people can execute fundamental cognitive processes. (p. 155)

Kail's working hypothesis is that there is a global processing speed factor that changes with age but is not linked to specific domains of information processing. Therefore, if RTs on various ECTs are limited by a global mechanism, the same form of growth curve in processing speed (or other RT parameters of RT) should occur across quite different ECTs. In fact, Kail (1991a,b) and Hale (1990) have demonstrated empirically that growth curves for the rates of processing speed on various tasks, such as memory search, visual search, mental rotation, and mental addition can be described by the same exponential equation,

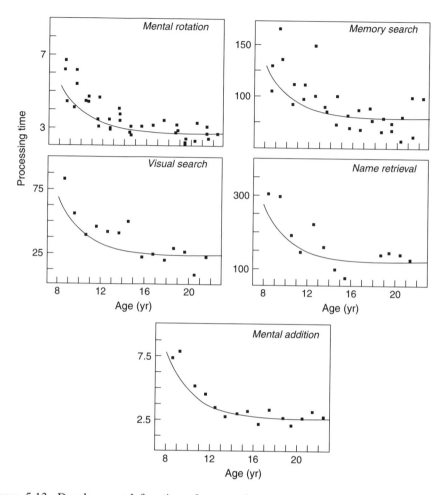

Figure 5.13: Developmental functions for mental processing time on various cognitive tasks. Each of the best-fitting exponential curves (solid line) for the data plots of the five tasks has the same decay parameter as a function of age. (Reprinted from Kail (1991b), with permission of Academic Press Inc.)

accounting for upward of 90 percent of the age-related variance in processing times, as shown in Figure 5.13.

RTs obtained on various paradigms and ECTs at a given preadult (P) age can be expressed as a multiplicative factor of the corresponding RTs obtained by adults (A). That is, $RT_P = mRT_A$, where $m$ is the multiplicative factor associated with age. For young adults $m = 1$; for preadults (and adults in later maturity) $m > 1$. Hence, if a given processing task requires the execution of three processes A, B, C, for example, which take the times $a$, $b$, and $c$, respectively, then the total RT will be $RT = (ma + mb + mc) = m(a + b + c)$. Kail's (1991a) meta-analysis of age changes in RT shows that the $m$ factor decreases exponentially between childhood and adulthood. Analogizing to computers, Kail (1991b) writes,

> If two computers have identical software, but one machine has a slower cycle time (i.e., the time for the central processor to execute a single instruction), that machine will execute all processes more slowly, by an amount that depends on the total number of instructions to be executed. . . . Speed of cognitive processing might be limited by the speed with which the human information processor can execute a fundamental cognitive instruction. (p. 179)

Accordingly, the course of mental development is largely accounted for by a growth curve of increasing speed of information processing, which is general across all cognitive tasks. This was nicely demonstrated in a study by Fry and Hale (1996), in which a variety of processing tasks were given to school children in Grades 2–7 and their RTs on the various tasks were compared with adults' RTs on the same tasks, with the results shown in Figure 5.14. Structural equation modeling showed that the developmental changes in RT strongly mediated improvements in working memory (WM), which is the link connecting RT to fluid intelligence, or psychometric *g*, as estimated in this study by Raven's standard progressive matrices (a nonverbal test of inductive and deductive reasoning ability). As Kail (1991b) has surmised, "[S]ome important qualitative-looking changes in cognitive development may be attributable, at least in part, to limitations in the speed with which

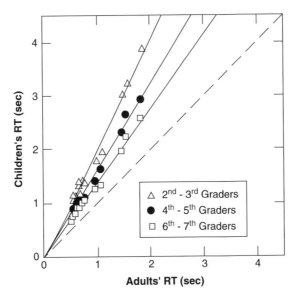

Figure 5.14: Mean RTs for Raven-matched groups of children as function of a matched adult group's mean RTs on a variety of RT paradigms and conditions (14 in all), showing that the various RTs in each age group (2nd, 3rd, and 4th graders) are regular multiplicative functions of the adult RTs (which would plot on the dashed line). (Reprinted from Fry and Hale (1996), with permission of the American Psychological Society.)

children process information. . . . The impact of processing speed on cognitive development may be widespread and may represent a fundamental mechanism of cognitive developmental change" (p. 182).

Although a global process shows up throughout these studies, it must also be pointed out that specific factors may also influence processing speed, causing the mean RT for certain domain-specific tasks to depart markedly from the general systematic regression of RT parameters on age. This independence of performance from the global process usually results from high levels of practice on a specific task, leading to the acquisition of task-specific strategies and the automatization of task performance, which reduces the role of controlled processing that depends on the cognitive resources that constitute the global speed factor in cognitive development.

The cause of the global mechanism in RT and its systematic change with maturation is not yet known. It plausibly involves the growth of brain structures and functions during childhood and adolescence. For example, EEG records taken at intervals over the developmental period show shifts from slower to faster brain waves, and the myelination of neural axons, which conduct the nerve impulses, gradually increases from infancy to early adulthood. Myelination increases neural-conduction velocity (NCV), which is found to be slower both in children and elderly adults. In fact, the trend of NCV as a function of age resembles the curves shown in Figures 5.1 and 5.2 at the beginning of this chapter. Brain research will eventually identify the physiological basis of the global factor in RT. It may well turn out to be the same as the basis of the highest common factor, $g$, of various psychometric tests of cognitive ability.

# Chapter 6

# Chronometry of Cognitive Aging

The gradual decline in mental functions that results with increasing age, beginning soon after maturity, has long been recognized. The whole scientific literature on cognitive aging is now much too vast to be summarized in this chapter; the subject has been most comprehensively reviewed in a book by Salthouse (1991). Also, Rabbitt (2002) provides an admirably detailed, analytical, and theoretical treatment of the subject, including sharply critical commentary on certain topics (particularly on the limitations of Brinley plots) that are prominent features of the present chapter, which deals only with the main findings and questions specific to the chronometry of cognitive aging.

Aside from casual observations dating back to antiquity, mental decline was first assessed quantitatively by means of conventional psychometric tests. In norming the first Wechsler–Bellevue test of adult intelligence on a fairly representative sample of the United States adult population, in which all 10 of the diverse verbal and nonverbal subtests were given to successive age groups from 18 to 70 years of age, David Wechsler (1944) noted the average decline in the tests' raw scores with increasing age. Further, as shown in Figure 6.1, he also pointed out a parallel loss in average brain weight, suggesting an organic basis for the decline in test scores. Of course, rates of decline in other physical attributes as well, such as vital capacity, visual and auditory acuity, and stature, parallel the decline in test scores. In observing these physical correlates of aging, Wechsler (1944) boldly wrote, "We have put forward the hypothesis that the decline of mental ability with age is part of the general organic process which constitutes the universal phenomenon of senescence, and have insisted upon the fact that the phenomenon begins relatively early in life. The evidence we have adduced for this hypothesis is the observed parallelism we have found in the rate of decline of various physical and mental abilities." (p. 59)

Ignoring the limitation that Wechsler's age data are cross-sectional rather than longitudinal, causal inference or detailed analysis from such correlations or graphical parallelism between psychological and physical variables, including the brain, remains conjectural without converging lines of evidence derived from other measurements as well, both behavioral and physical. Such evidence is obtained nowadays with the methods of psychometrics, experimental psychology, electrophysiology, and brain imaging. Modern research clearly supports Wechsler's hypothesis (Bigler, Johnson, Jackson, & Blatter, 1995; Deary & Der, 2005a, b). It might also be noticed that in Figure 6.1 the appearance of parallelism could itself be merely an artifact, as the scores representing intelligence, unlike the equal interval and ratio scale measurement of brain weight, allows no stronger inference than is permitted by an ordinal scale: the mental decline *per se* is not in question, but the shape of the curve across age levels has no true meaning. Meaningful comparisons of the shapes of the physical and mental decline curves can be made only if both variables are measured on a ratio (or at least equal-interval) scale. Hence mental chronometry is increasingly appreciated among investigative techniques in recent research on cognitive aging.

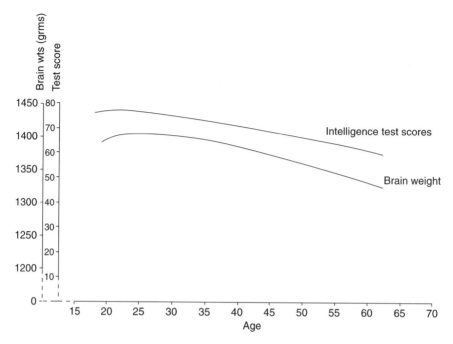

Figure 6.1: Curves showing both the mean raw scores on the Wechsler–Bellevue Adult Intelligence Scale and the mean weights of autopsied brains as a function of age in years. (Reprinted from Wechsler, 1944, with permission of The Williams & Wilkins Company.)

The increasing percentage of the population that survives into the seventh, eighth, and later decades, as well as the health problems associated with this age group, has spurred great interest and research support for the scientific study of cognitive aging. The tools of mental chronometry are playing a prominent part and will continue to do so to an even greater extent as causal connections are sought between behavioral variables and brain mechanisms. The main reasons for the usefulness of chronometry are not only the advantages of its absolute scale properties, but also its sensitivity and precision for measuring small changes in cognitive functioning, the unlimited repeatability of measurements under identical procedures, the adaptability of chronometric techniques for measuring a variety of cognitive processes, and the possibility of obtaining the same measurements with consistently identical tasks and procedures over an extremely wide age range. These properties are possessed by hardly any conventional psychometric tests. Timothy A. Salthouse (1996), a leading researcher on cognitive aging, has stated well the over-arching advantage of chronometrics in the study of inter- and intraindividual variation in cognitive functions:

> Speed may function as a bridging construct between behavioral and neuro-physiological research. Because time is an objective and absolute dimension rather than a norm-reference scale, as is the case with most behavioral measures, it is inherently meaningful in all disciplines and this has the

potential to function as a Rosetta stone in linking concepts from different disciplines. (p. 425)

The overall picture of mental decline, beginning in the mid-1920s, as shown by chronometric tests, is much like a mirror image of the developmental curves from early childhood to maturity, as shown in Chapter 5 (Figures 5.1 and 5.2). The extent to which similar processes are involved in both sides of this almost symmetrical curve of growth and decline is one of the main questions for empirical inquiry. A leading question in research on cognitive aging is whether the observed decline in performance is mostly due to changes in strategic processes or to nonstrategic processes. In factor analytic terms, do the changes with age reside mostly in a large number of specific cognitive abilities or do they reside overwhelmingly in just one general factor? Is response time (RT) variance associated with advancing age highly varied and narrowly localized in different functions, or is it mostly global across all mental functions?

Experimental psychologists have typically eschewed invoking latent variables in their explanations of behavior. Particularly troubling for the experimenter is the fact of large individual differences in performance under identical conditions of the experimental task. Behavioral variables that cannot be directly manipulated by the experimenter traditionally have been of little interest in their own right. The experimental psychology of aging, however, has now virtually forced us to pay attention to latent variables.

Individual variability in behavior has traditionally been the province of psychometrics and differential psychology, in which explanations and formulations are expressed in terms of factors, latent variables, or hypothetical constructs, which cannot be observed or measured directly but can only be inferred from correlations among various types of cognitive performance that evince common factors. But rigorous causal explanations can rarely be made solely from correlation analysis. The critical step for explanation is to hypothesize structural and physiological changes in the brain to account for the latent variables.

Such hypothesizing comes cheap. The crucial step is for the brain–behavior connection to be demonstrated empirically. This calls for the methodology of the brain sciences. The explanation of a psychological phenomenon or of individual differences therein is sought, wholly or in part, in terms of variation in certain anatomical structures and physiological processes in the brain. Such "reductionism," however, does not in the least obviate the need for behavioral measurements and their correlational analysis. The precise measurement of individual variability in cognitive phenomena *per se*, as represented by well-defined types, processes, factors, or theoretical constructs, is necessarily an integral part of discovering and describing its connection with the brain's physical structures and processes. The investigation of behavior–brain connections is only possible in terms of precisely measured variables in both the physical and behavioral domains.

On the behavioral side, chronometry serves this purpose exceptionally well. And, as indicated by brain wave measurements, the brain itself is inherently a kind of time machine or biological clock, with important aspects of its functions reflecting continuous and fluctuating neural activity that is measurable in real time.

Before describing the most typical findings on the chronometry of cognitive aging, however, readers should be reminded of a number of methodological caveats and conditions that pertain generally to this class of studies.

1. *Selection bias in subject samples.* Most of the aging research is based on cross-sectional samples. Different age cohorts in a cross-sectional sample, therefore, may differ systematically in many other ways besides age, such as education and socio-economic status, intelligence, childhood diseases, and medical history. In many studies, some of these factors are controlled to an extent by selection and matching different age cohorts to minimize variance on possibly confounding variables, most often education and general ability.

Successively older age cohorts, however, are increasingly liable to the so-called *terminal drop.* This is the relatively sudden and rapid decline in mental test scores during the interval of from several months to a year preceding the individual's death. Longitudinal studies, in which the phenomenon was first discovered, as well as cross-sectional studies, are liable to this increasing source of variance in the older cohorts. The terminal drop tends to lower the mean performance level of an age cohort more than would be expected if none of the subjects showed the terminal drop and it does so disproportionally in the older age cohorts; it therefore contributes a possibly unwanted source of variance that may not be intrinsic to the normal aging process prior to the terminal drop. This effect can be controlled to some extent only retrospectively, separating out those members of a given age cohort who died within a year after being tested. Few, if any, of the studies reviewed here, however, have controlled for the terminal drop.

A method I have not seen used in the literature on aging is the sibling design, although it has been used to study developmental changes in IQ (Jensen, 1977). While actually being cross-sectional, it has some of the advantages of a longitudinal study. It controls to some extent for family background, education, life style, and medical status. The method consists simply of placing singleton siblings into different narrow-band age cohorts. Naturally, wide age differences are seldom spanned by siblings; most siblings can only be entered into cohorts separated by a few years. The sibling correlations on whatever psychological measures are obtained can also be scientifically informative, as I have explained elsewhere (Jensen, 1980a). In the same way, young adults and their parents can also be used to create cross-sectional cohorts. There is no reason to expect the *average* difference between different aged groups of adult siblings on any mental tests to be different from zero except for factors attributable to their age difference. A birth order effect on RT is as yet unknown, and this would have to be empirically determined.

Another aspect of subject selection to be kept in mind is that most studies of normal cognitive aging are based on samples of self-selected volunteers. When we have solicited elderly volunteers for testing through notices in local newspapers, for example, we have been impressed by the general level of physical and mental fitness of those who choose to participate. Direct personal appeals for volunteers living in retirement homes and the like yield only a small minority of the residents; the vast majority cannot be persuaded to participate in a study. Observing many individuals over the age of 65 who refuse to volunteer, one is impressed by their noticeable differences in vitality from those who willingly come to the laboratory. It is likely that most studies of cognitive aging that are based largely on volunteer samples, are not ideally representative of the total aging population. What one sees in the reported research, I suspect, represents probably only the more fortunate half of the aging population. Even so, as will be seen, the effects of aging are clearly apparent in their chronometric measurements.

The extrinsic factors that may be correlated with age differences are of little concern in clinical studies of any single individual, whether in assessing cognitive changes with age, medications, or some progressive pathology. The crucial requirement in this instance is the initial establishment of the given individual's baseline measurements on a variety of chronometric tasks against which later changes can be reliably evaluated. In addition, consecutive day-to-day measures of within-subject variability in any particular test should be obtained as a basis for estimating the probable error of measurement for the given individual. This statistic is necessary for interpreting future deviations from the individual's earlier baseline. That such repeated testing under identical conditions is possible with chronometric tasks is one of their chief advantages over conventional psychometric tests.

2. *Physical demands of the test.* Physical demands of the test should be kept at a minimum in research on aging, so as not to confound cognitive changes with possibly extrinsic sensory-motor limitations. These take mainly two forms: endurance and dexterity. In a repetitive but attention-demanding task, elderly subjects show fatigue sooner than younger subjects. This can be largely mitigated by the experimenter's close observation of the subject's on-going performance and by appropriate rest intervals and changes in activity. Students in high school and especially college students best tolerate speeded tasks for quite long sessions without showing diminished performance. Regardless of the subject's age, the tester's aim should be to elicit the subject's best performance, even if fewer or shorter tests have to be given during a session.

The motor demands of the response are most important in testing the elderly, and should be kept as simple and as uniform as possible for all tasks. This particularly involves the nature of the response console. We have found that consoles consisting of keyboards on which subjects must operate the keys by placing two or more fingers of one or both hands simultaneously are the least satisfactory. It requires more initial learning and practice trials, and is more susceptible to any conditions that affect fine motor control and dexterity, such as tremor or arthritic limitations.

By far the best arrangement, especially for children and the elderly, is a response console consisting of a home key which the subject presses with one finger and releases by lifting the finger in response to the reaction stimulus (RS), then depresses another key a short distance from the home key. (Such a console is shown in Figure 2.4). A *lift* response is less liable to tremor or other motor limitations than a *press* response, and the time of the whole operation in a given trial is divided into two components: RT (i.e., the interval between the RS and lift) and the *movement time* (MT, i.e., the interval between *lift* and *press*). The response keys or buttons should be sufficiently large, so that aiming dexterity and precision are minimized in RT and MT.

Moreover, one and the same RT–MT response console can be adapted to diverse paradigms, with the stimulus conditions usually presented on a computer monitor. Familiarity with a single mode of responding in every paradigm is a marked advantage in reducing the number of training trials, making all tasks easier, and, above all, controlling variation in the motor-skill components of RTs across different cognitive tasks. It is seldom possible to make meaningful direct comparisons of the absolute RTs obtained on different paradigms when they are obtained on different response consoles. The marked variations in apparatus and procedures used in different laboratories, even for nominally the same paradigm, limit the discovery of certain regularities within and across paradigms and the interpretation of absolute RT differences between paradigms. Time is indeed a universally valid and absolute unit of measurement, but this scientific advantage is sadly diminished for the behavioral sciences to the extent that the conditions under which time intervals are measured lack standardization. Literature reviews and secondary analyses of reported data in this field are presently disadvantaged by this limitation.

3. *Confusing psychometric and chronometric measures of speed.* Many psychometric tests, such as the Wechsler scales and the Armed Services Vocational Aptitude Battery, contain certain timed subtests that are given with instructions emphasizing speed and accuracy. All of the individual items are typically very easy, so errors are almost nonexistent, and the subject's score is the number of error-free items completed within a specified time limit. Examples are clerical checking of pairs of four-digit numbers as being either the same or different, consistently canceling or ticking a specified letter in a long series of random letters, or adding pairs of single-digit numbers. In hierarchical factor analyses of many psychometric tests, these kinds of speed tests show up usually only as a first-order factor, with little of their variance showing up in any higher-order factors, and they have the smallest *g* loadings of any types of items included in mental abilities test batteries. Their loadings on higher-order factors, however, increase somewhat in older age groups. Nevertheless, I have omitted these kinds of speed tests from consideration here, because there is evidence that they reflect something quite profoundly different from the RTs measured in the various well-known chronometric paradigms.

Surprisingly, of all the subtests found in standard psychometric batteries, the clerical speed tests of the types previously described have *lower* correlations than do any nonspeeded tests with the RTs measured by true chronometric tasks. Moreover, the degree to which these chronometric RTs are correlated with various nonspeeded psychometric tests (e.g., vocabulary, comprehension, general information, logical reasoning) is positively related to these tests' loadings on psychometric *g*.

In contrast, chronometric tests, in which the RT is measured separately for each response unit and in which sensory-motor skills are minimized, apparently tap a deeper, more intensely cognitive level of information processing than do the speeded tests of conventional psychometric batteries.

4. *RT specificity and the importance of aggregation and factor analysis.* It is easy to forget that the reliability and validity of psychometric tests depend heavily on the aggregation of many different items. Each test item is loaded with a remarkably small *signal* (reflecting the *true score* derived from the whole test) and a great deal of *noise* (reflecting the item's *specificity* and *measurement error*). Only by aggregating a large number of single-item scores into a total score can the test attain sufficiently high reliability and external validity to be of any practical or scientific value. The same principle applies even more strongly to RT tests because of an important difference between psychometric and chronometric tests. The RTs measured on every trial in any given RT paradigm are so much more homogeneous than are psychometric test items, which are relatively diverse, that the aggregation of RTs over *n* trials yields a true score measure of the individual's mean or median RT which represents a much narrower factor than is the true score of a conventional test composed of *n* diverse items. Although the internal consistency

reliability of the composite RT obtained from any particular paradigm can be very high, its *specificity* when factor analyzed among other RT paradigms or external variables is also very large because of the homogeneity of the RT data obtained in any given paradigm. This condition has to be taken into consideration when looking at the correlations of RT with various external variables, such as age and test scores. The RT correlations with external variables, though statistically significant, is typically small. As a rule, however, when RT scores from two or more paradigms are aggregated, either by summation or multiple regression, the RT correlation with external variables is markedly increased. The correlation of single-paradigm RTs with IQ, for example, are typically in the .30's, whereas the correlation based on an aggregate of RTs obtained in different paradigms increases up to as large as .70 or above with the inclusion of more different paradigms in the composite score.

Another point to consider is that in RT measurement, *internal consistency reliability* is an ambiguous if not misleading concept. What is usually estimated as the error component in a subject's overall RT (i.e., the root mean square deviations of the subject's single trial RTs from the subject's overall mean) is none other than the SD of the subject's RTs, labeled reaction time standard deviation (RTSD). But RTSD is itself a reliable parameter of the subject's chronometric performance, having substantial correlations with external variables such as age and IQ. RTSD is also intrinsically related to the individual's mean RT, the two measures being almost perfectly correlated in large population samples. In other words, trial-to-trial deviations in RT from the subject's mean RT over a number of trials cannot be considered experimental or random error variance in the usual sense of the term. It is simply intraindividual variability, which is a reliably measurable characteristic in its own right and is inextricably linked to the subject's mean RT. However, one can estimate the average *consistency* with which the subjects in a given sample maintain the same relative positions in RTs among other subjects across a given number of trials by means of a two-way analysis of variance (Trials × Subjects) in which $Consistency = (F_S - 1)/F_S$, where $F_S$ is the (Between subjects MS)/(Residual MS). This *consistency coefficient* is analogous to a reliability coefficient without implying measurement error in the usual sense. It measures the consistency of individual differences in RT across trials.

Test–retest reliability of the subjects' mean RTs is unambiguous, being a measure of the subjects' stability across different test sessions separated by a specified time interval. Fluctuation of mean RT across sessions separated by hours or days or weeks is a normal condition and can have many causes. Nevertheless, relative to true score individual differences, such fluctuations are relatively minor, as shown by the satisfactory test–retest reliability of subjects' mean RTs obtained with a particular RT task.

Common factor analysis (and principal components analysis) of a number of RT tests provides probably the best method for separating the specificity of various RT tasks from their commonality. Regularities and lawful relationships are often obscured by the raw measurements from any particular RT test. This can be overcome by simply combining the RT data from a number of tests, or by obtaining factor scores, which are factor-weighted linear composites of the various test scores. The advantage of factor scores (particularly if they are based on a principal components analysis) is that they represent the independent additive components of the total variance in the whole battery of different

RT measures. The largest common factor (or first principal component, PC1) of a number of RT tests typically shows the largest correlation with external variables. The large amount of specific variance in any particular RT tests, regardless of how many trials are given, greatly constrains its correlation with other variables. The relatively global measures of RT achieved by aggregation or by factor analysis generally show more lawful and theoretically interesting relationships with age or other variables.

Aggregation of data *across subjects* also can serve a similar purpose as aggregation across tests, revealing trends or regularities in the data that might otherwise be obscured by the welter of individual variation. Subjects with similar values on variable *X* (e.g., age) are aggregated into *N* groups and these groups' means on variable *Y* (e.g., RT) are then plotted as a function of the group means on *X*. The correlation coefficient $r_{XY}$ obtained from such a scatter diagram is called an *ecological* correlation; it is most commonly used in epidemiology but can also be useful in psychology (Lubinski & Humphreys, 1996). A form of ecological correlation or regression called a Brinley plot has been a useful method in studies of cognitive aging, as graphically demonstrated later on.

5. *Relatively specific but important RT factors can be obscured by global factors.* This is the other side of the issue discussed above and reinforces the argument for extracting the largest common factor from a battery of RT tests so that the remaining significant factors (or principal components with eigenvalues greater than 1) can be identified and measured for their diagnostic significance or relationship to other variables independently of the large global factor in RT tests. For similar reasons astronomical observatories are located far away from city lights to obtain clearer images of the most distant objects in the night sky. The factor scores based on the most general factor in a battery of RT tests (e.g., PC1) can be regressed out of the raw scores on the various RT paradigms so that these regressed scores then can be studied in relation to external variables of interest, such as age, brain pathologies, drug effects, special cognitive abilities and the like. The independent contributions of various RT paradigms can thus be more readily discerned and their external validity more finely pinpointed in relation to specific cognitive processes than is easily possible when RT data from a number of different paradigms all share the same global factor. It is likely that a specific paradigm has significant external validity for a particular purpose over and above any contribution of the global factor it has in common with other paradigms.

6. *Correlational weakness of difference scores and slope measures.* Certain analyses of RT data make use of *difference scores* and *slope measures*, whose statistical limitations are similar in that they both compound measurement error, thereby often drastically attenuating their correlations with external variables. Two prominent but simple examples are the Posner and the Hick paradigms. The Posner paradigm is a binary RT task based on the difference in RT between two conditions, labeled *physical identity* (PI) and *name identity* (NI). PI is the RT for recognizing whether two letters (or two words) presented simultaneously are *physically* the *same* (e.g., **A A**) or different (e.g., **A a**). NI is the RT for recognizing whether they have the same *names* (e.g., **A a**) or different names (e.g., **A B**). Theoretical interest is in the difference between the mean RTs for NI and PI, i.e., NI−PI, which is always a positive value, because NI requires more steps in processing, having to access the well-learned codes from long-term memory (LTM), whereas PI processing involves a perceptual discrimination but not semantic

information that has to be accessed in LTM. Hence the NI−PI difference between the mean RTs estimates the time needed for accessing and retrieving letter codes in LTM. (NI−PI is also measured using highly familiar words, with the NI condition consisting of synonyms or antonyms.)

So far so good. The problem arises, however, when the NI−PI difference scores are correlated with an external variable hypothesized to be related to retrieval of verbal information from LTM, such as scores on tests of vocabulary and verbal analogies. The culprit is the low reliability of the NI−PI value, which is a difference score, and difference scores are always much less reliable than are the separate component scores when the components are highly correlated with each other. Because the mean RTs for NI and PI are quite highly correlated with each other, the reliability of the NI−PI difference is markedly reduced. And of course low reliability of a measure limits its possible correlation with any other variable. The same caution applies to all other examples of the subtraction method introduced by Donders, such as the difference between choice RT and simple RT. The reliability of a difference score $d$, where $d = X - Y$, is $r_d = (r_{XX} + r_{YY} - 2r_{XY})/(2 - 2r_{XY})$. [*Note*: The reliability of the *sum* of two scores is the same as the above formula with the minus signs changed to plus signs.] If one has highly reliable measures of $r_{XX}$ and $r_{YY}$, then $r_d$ can be used in the well-known Spearman formula to correct for attenuation of the correlation of $d$ with any other variable. It should be noted that the problem is greatly lessened when looking at the difference between raw score group means on variables $X$ and $Y$ because the variance due to random errors of measurement tends to average out in the mean by a factor of $1/N$ as a function of sample size ($N$). However, statistical tests of the significance of the difference between means are affected by measurement error, as it enters into the SD of the scores and thence into the standard error of the means and the SE of the mean difference.

The Hick (H) paradigm illustrates the statistical problems with any RT data in which individuals show systematically increasing (or deceasing) RTs and these trends are measured as the slope parameter of a linear regression equation. In the Hick paradigm, for example, there is a theoretical interest in the slope (or linear increase) in RT as a function of task complexity or information load of the RS as measured in BITs (binary digits). In the regression of RT on the number of BITs, the regression constant, or intercept, captures the general factor of individual variation whereas the slope represents the residual variance, which is a relatively small proportion of the total variance of individual differences. Moreover, because the intercept and slope share the same errors of measurement for any given individual, the intercept and slope are always negatively correlated, as I have explicated in detail elsewhere (Jensen, 1987a, pp. 142–146, 1998b). These conditions play havoc with the correlation between individual differences in the slope and any external variable, severely attenuating the true correlation. The intercept acts as a suppressor variable in the correlation of the slope parameter with any other variable; therefore, using partial correlation to remove the influence of the intercept is preferable to a simple (zero-order) correlation. The most suitable way to show the theoretically predicted relationship between slope and some other variable (e.g., psychometric $g$) is by comparing the slopes of the *means* of two or more groups of subjects that differ discretely on the external variable (Jensen, 1998a, 1998b).

### *Renaming a Phenomenon Is Not a Causal Explanation of It*

One must beware of the common tendency to explain a psychological phenomenon simply in terms that merely rename the observed phenomenon. The commonest example of this in the chronometric literature is in explaining RT phenomena, particularly individual and age differences in RT, as resulting from differences in "attentional resources." Usually this means no more than the RT phenomena observed in the first place, or whatever other variable is correlated with RT, such as psychometric *g*, as if we already know more about the nature of "attentional resources" than we know about the RT phenomena it is supposed to explain. Unless a construct can be defined and measured independently of the RT phenomenon under investigation, it can contribute nothing to our understanding of the phenomenon.

We are more apt to derive potentially productive causal hypotheses from multivariate analyses of a correlational nexus in which the RT phenomenon of interest is included among other operationally defined and precisely measured variables. The presumed causal pathways between independent, operationally defined constructs derived through structural equation models provide a source of hypotheses for testing causal explanations at a physiological level. Physiological processes are more likely to be reflected by the common factor among a number of interrelated behavioral measures than by any one of them alone. Factor specificity in RT data is more apt to reflect complex interactions of apparatus and procedural variables than central processing variables. Research on cognitive aging, for example, has rather consistently turned up the relationships, derived from path analysis or fitting data to structural equation models, shown in Figure 6.2. It indicates that age affects a general speed factor. This has direct effects on both memory and psychometric *g*. And the memory factor also has an effect on *g* independent of mental speed. Thus it appears, at present, that there are at least two independent factors involved in normal cognitive aging: mental *speed* and recall *memory*, the speed factor being the more general as shown in the path diagram in Figure 6.2. An empirically confirmed corollary is that the correlation between mental speed and general intelligence (*g*) is only slightly reduced when age is partialed out, whereas the correlation between age and *g* is reduced to nearly zero when mental speed is partialed out (Bors & Forrin, 1995).

In the elderly, there is also an increase in the number and the magnitude of correlations of RT with different physical variables in which there is age-related decline (e.g., sight, hearing, muscular strength). But there is no direct causal connection between these peripheral

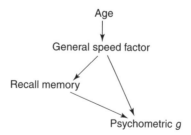

Figure 6.2: A best-fitting path diagram of the hypothesized causal relationships between age and the constructs of mental speed, recall memory, and psychometric *g*.

variables and RT. This is shown by the fact that when age is statistically controlled by partial correlation these peripheral variables' correlation with RT is reduced to nearly zero.

## Age Effects in Standard Reaction Time Paradigms

Besides the generalization that various measures of RT increase with age in later maturity, two other generalizations are commonly seen in the RT literature:

(1) Younger–older age differences in RT are an increasing function of task complexity (i.e., the information processing load).
(2) RT increases as a function of the amount of information that has to be retained to complete the task or the task's demand on what is referred to as the *working memory* (WM).

The latter condition is typically controlled experimentally by means of a *dual* RT paradigm, in which the subject must hold in memory the reaction stimulus ($RS_1$) of the first paradigm ($RT_1$) while responding to the $RS_2$ of the second paradigm ($RT_2$). For quite some time, it was believed that the effects of increasing task complexity and of dual tasks might be some exponential function of increasing age; that is, the increase (slowing) in RT is an accelerated function of increasing age, and that different paradigms probably show quite different exponential curves, depending on the particular cognitive processes presumably involved in a given RT paradigm. In recent years research aimed at this question has provided a quite clear answer as demonstrated in the following examples:

### *Analysis of RT Paradigms in Young and Elderly Groups*

The kinds of cognitive aging variables that are typically examined in RT research can be most simply introduced by studies of widely separated age groups tested on several standard RT paradigms and dual tasks. It is advantageous to begin with two matching studies performed in my laboratory at Berkeley (Ananda, 1985; Vernon, 1981a, 1983). They permit exact comparisons of younger and older age groups of comparable educational level. The various test paradigms and procedures were carried out under virtually identical conditions — the room, test apparatus, instructions, and practice trials were the same in both studies.

The eight RT paradigms in these studies are as follows: (in all the cases RT is the time interval between the onset of the RS and the subject's lifting the index finger from the Home key.)

**Hick paradigm**   This uses the Jensen box with RTs measured separately for 1, 2, 4, and 8 response alternatives (i.e., 0, 1, 2, 3 BITs), respectively (see Figure 2.10).

**Sternberg memory scan paradigm**   This uses a computer monitor and a binary response console (see Figure 2.4). Test trials presented digit strings of randomly varying length from 1 to 7 nonrepeated random digits (selected from 1 to 9), with 84 different strings in all (hence 84 test trials). On a random half of the trials the probe digit was a member of the positive set (subject presses the button labeled YES) and on half the trials the probe digit was not in the positive set (subject presses the button labeled NO).

**Same–Different paradigm (S–D)**   This is one element (PI) in a more complex version of the classic *Posner paradigm* (see p. 20). It uses a computer monitor and a binary response console. A pair of words is presented simultaneously. They are either physically identical (e.g., CAT–CAT) or physically different (e.g., CAT–MAT). The subject responds by pressing one of the two buttons labeled Same or Different. Immediately, informative feedback, Correct or Incorrect, appears on the computer screen. Note that in this task it is not necessary to access these words in memory; the difference is purely perceptual rather than semantic.

**Synonyms–Antonyms paradigm (S–A)**   This is the other element (Name–Identity) in the complex version of the Posner paradigm. It is set up in the same way as the S–D paradigm, except that the paired words are either synonyms (*quick–fast*) or antonyms (*sharp–dull*). The subject responds by pressing buttons labeled Same or Different, which is followed by informative feedback. Note that the meanings of the words have to be accessed in LTM, which is an additional, more complex process than the perceptual task in the S–D paradigm and therefore has a longer RT.

**Dual tasks**   These consist of combinations of the MS, S–D, and S–A tests described above. In each case a MS digit series is presented first, but the probe digit and the subject's response to it are held in abeyance while either the S–D or the S–A task is presented, after which the probe digit appears and the subject responds YES or NO as in the simple MS task. Thus two RTs ($RT_1$ and $RT_2$) are measured in each dual task: $RT_1$ for the MS and $RT_2$ for either the S–D or the S–A task. The time sequence for the dual task in which S–D is interposed, labeled DT:MS(S–D), was as follows:

| Event | Time |
|---|---|
| Preparatory signal ("beep") | 1 s |
| Blank screen | 1 s |
| **48351** (digit set appears) | 2 s |
| Blank screen (random interval) | 2–4 s |
| **HOT–DOT** (word pair appears) | |
| Subject presses button **Different** | $RT_1$ and $MT_1$ are recorded |
| Blank screen | 1 s |
| **3** (probe digit appears) | |
| Subject presses button **YES** | $RT_2$ and $MT_2$ are recorded |

   In all there are 11 distinct and experimentally independent types of RT measures, each coded (in brackets) as follows:
*Single RT measures*

1–4.  Hick RT for 0, 1, 2, 3 BITs (i.e., $H_0$, $H_1$, $H_2$, $H_3$, or H mean)
   5.  Digit memory scan (MS)
   6.  Words: physically Same–Different (S–D)
   7.  Words: semantically Synonym–Antonym (S–A)

*Dual task RT measures*

8. Dual task: MS with S–D interposed [DT:MS(S–D)]
9. Dual task: MS with S–A interposed [DT:MS(S–A)]
10. S–D as interposed task in dual MS task [DT:S–D]
11. S–A as interposed task in dual MS task [DT:S–A]

The Young group in this study comprised 100 volunteer university students aged 18–34 and a mean age of 21.4 years (SD 2.63). The Old group were 67 volunteer adults aged 51–87 and a mean age of 67.8 years (SD 8.65). Their mean years of education was 15.3 years. They were reportedly in good health, free of visual or motor impairments that would handicap their performance, and appeared exceptionally fit for their age.

Although MT was measured on all tasks, it is not considered further here, because it is scarcely related to the cognitive complexity of the various tasks. While the various tasks show large mean differences in RT, variation in MT is remarkably small across the different tasks within either age group. Figure 6.3 compares the mean response latencies for RT and MT across the various tasks in the Old group. The overall mean difference in MT between the Old and Young groups is 78 ms as compared with their overall difference of 190 ms in RT. The Old and Young mean MTs are correlated +.65 across the 11 tasks as compared with +.99 for the RTs. This aspect of RT in relation to age, as shown in the following analyses, is of major theoretical importance.

**Brinley plots**    A graphically revealing method for comparing groups simultaneously on a variety of RT tasks was originally suggested by Joseph F. Brinley (1965). It consists of plotting the various RTs for one group as a function of the RTs for the other group. This

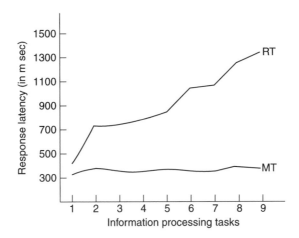

Figure 6.3: Mean RT and M in the Old group. Task Number 1 is the Hick mean. (From Ananda, 1985). *Note*: The numbers on the abscissa represent the rank order of magnitude of the RT on each of the tasks. Unlike RT, MT is virtually unrelated to the varying cognitive demands of the tasks.

method, termed a "Brinley plot" quite clearly reveals the form of the relationship between the compared groups and is used in many of the following analyses.

**Hick paradigm**   The simplicity and homogeneity of task demands across the four conditions (0–3 BITs) in this paradigm make it an ideal test of age differences within a relatively narrow range of RT. The Brinley plot reveals no significant departure from linearity of the regression of the Old groups' mean RTs for 0, 1, 2, 3 BITs on the Young group's mean RTs, as shown in Figure 6.4, indicating that the mean RT's of the Old group are predicted almost perfectly ($r = .996$) by the mean RTs of the Young group. The differences in mean RT between the groups increase as a linearly increasing function of task complexity scaled in BITs. Increasing task complexity in this case simply multiplies the Young group's RT by a constant factor to estimate the Old group's RT at each greater level of complexity. But of course the Hick paradigm is very homogeneous across every level of complexity, from 0 to 3 BITs, and these might all involve the very same processes, though to different degrees depending on the information load.

But what if we make a Brinley plot of mean RTs on the diverse RT paradigms described above, with different paradigms clearly making different demands on stimulus apprehension, perceptual discrimination, storage and retrieval of information from short-term memory (STM), and dual task conditions of storage/processing trade-off involving working memory capacity? Do the different cognitive task demands of these varied tasks interact differentially with age? Figure 6.5 shows the Brinley plot for all 11 of the RT data points. It is seen that the mean RTs of the Old group are highly predictable from the mean RTs of the Young group, suggesting that a single global multiplicative factor determines the increase in RT with age regardless of the specific cognitive demands of the task. The difference of

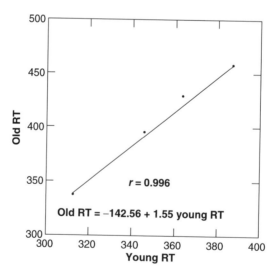

Figure 6.4:  A Brinley plot of mean RT for Old and Young subjects on the Hick paradigm. The data points from lowest to highest are for 1, 2, 4, and 8 response alternatives, representing 0, 1, 2, 3 BITs.

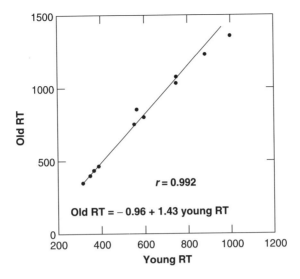

Figure 6.5: A Brinley plot of mean RT for Young and Old groups on eleven measures; the rank order of RTs is the same for both groups. [The RT tasks going from lowest to highest RT (indicated by the number of RT tasks listed on p. 106): Hick 1, 2, 3, 4, 5, 9, 8, 10, 6, 11, 7.]

the correlation of .992 from unity is surely within the margin of error. Note that the RTs of both the dual tasks and the single tasks fall on the same regression line, indicating the lack of interaction between age and the dual/single task demands, a finding that is corroborated in a study by Salthouse, Fristoe, Lineweaver, and Coon (1995). In the same data set as Figure 6.5, we can also look at the Brinley plot of the mean intraindividual variability (SDRT), shown in Figure 6.6, which reveals essentially the same global phenomenon.

Because the mean RTs of the Old group are predicted for various tasks by some constant multiplier of the Young group's RTs on the same tasks, it follows that the size of the RT difference between the Old and Young groups increases across tasks as the overall mean RT on those tasks increases. In terms of effect size (i.e., ES = mean difference/standard deviation), the Old–Young differences in RT are very large, with ES ranging between 1.1 and 1.8 across tasks. The relative sizes of the mean RTs on various chronometric tasks are often regarded as a measure of task complexity, conceived as the number of steps in information processing needed to complete the task successfully. For the tasks considered here, the best estimate, therefore, should be the mean RTs for each task averaged over both age groups. The result, shown in Figure 6.7, indicates a strong linear relationship ($r = .92$) between the age differences on various tasks and task complexity (as reflected by their mean RT).

The foregoing results again favor the hypothesis of a large general factor common to all the RT tasks in both age groups, causing the pattern of RTs across the various tasks to be almost identical despite the large absolute average difference in RTs between the Old and Young groups. A principal components (PC) analysis of the intercorrelations among the

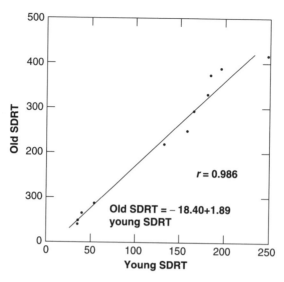

Figure 6.6: A Brinley plot of the mean intraindividual variability of RT (SDRT) in Young and Old groups for the 11 RT tasks shown in Figure 6.5. The rank order of the variables, from smallest to largest RTSD, is the same as those in Figure 6.5.

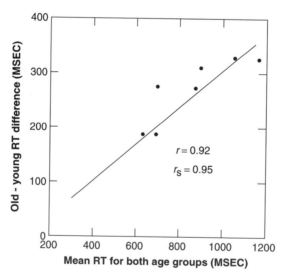

Figure 6.7: The difference between Young and Old RT plotted as a function of task complexity as indicated by the mean RT for both groups on each of the eight tasks, and the Pearson $r$ and Spearman rank correlation $r_S$ between the means and differences. Regression equation: Old–Young difference $= -23.86 + 0.335 \times$ (mean RT). [In rank order of magnitude from lowest to highest mean RT the tasks are (as numbered on pp. 106–107): Hick mean, 5, 9, 8, 10, 6, 11, 8.]

eight tasks in Figure 6.7 is probably the simplest way to represent the magnitude of this general factor. Only the PC1 of the correlation matrix has an eigenvalue greater than 1, and the residual correlation matrix is wholly nonsignificant. Essentially, the same phenomenon of a single homogeneous global factor was found in a principal components analysis of RTs from seven diverse information-processing tasks (Hale & Jansen, 1994; Myerson, Zheng, Hale, Jenkins, & Widaman, 2003). The PC1 in the present analyses accounts for 70.5% of the variance in all the tasks in the Young groups and 63.8% of the variance in the Old group; the difference of 6.7% is nonsignificant. Except for the mean RT on the Hick paradigm, with PC1 loadings of .52 and .47, in the Young and Old groups, respectively, all the other tasks had relatively homogeneous PC1 loadings ranging between .77 and .92. The PC1 vectors of the Young and Old groups are correlated .901 and their coefficient of congruence is .998. They could hardly be more alike.

A theoretically interesting question is the relation between the average loadings on the general factor (PC1) for the various tasks and the size of the Old–Young RT differences on those tasks. This correlation is $r = .803$, $p = .016$, indicating that the size of the Old–Young RT differences on these tasks are mainly associated with their one common factor. *Ipso facto*, the constant multiplier factor by which Young RTs predict Old RTs is apparently saturated with the general mental speed factor common to all of the diverse RT tasks. One may wonder whether this common factor in RT tasks, call it RT$g$, is the same as the psychometric $g$ of IQ tests. The present evidence is not conclusive, but strongly suggests that the general mental-speed factor associated with aging and psychometric $g$ are closely related. Vernon (1989a) has reported correlations between RT$g$ and psychometric $g$ in the range of |0.4|–|0.7| in several studies. Also, Nettelbeck and Rabbitt (1992) performed a study of 104 individuals, aged 54–85 years, in which they did a hierarchical factor analysis of a correlation matrix containing 15 variables: two content-free mental speed measures, inspection time (IT) and 4-choice RT (CRT); 11 conventional psychometric tests; and age. A second-order general factor extracted from this matrix and accounting for 40.7 percent of the total variance, had loadings on IT, CRT, and age of $-.71$, $-.69$, and $-.43$, respectively. None of these three variables had as large loadings on any of the three residualized first-order factors, although seven of the nonspeeded psychometric tests had substantial loadings on one or another of the first-order factors, and age had a loading of $-.39$ on a Memory factor (defined by learning and free recall of familiar words). The implication of such findings is well stated by Nettelbeck and Rabbitt (1992):

> [A]ge changes in all cognitive skills proceed at a common rate because they are all locked to age declines in information-processing rate. In other words, slowing of information processing, which relates to changes in the central nervous system and is not attributable to peripheral sensory or motor impairment, is a single master factor underlying involutional changes in all cognitive skills, including performance on IQ tests. (p. 191)

**Sternberg's memory scanning paradigm**   We should zero in on this particular paradigm, which is homogeneous in the type of its cognitive demands over a wide range of task difficulty, requiring the retrieval of simple information from STM. The task demands differ only quantitatively, according to the number of items (digits or letters) that have to be visually processed into STM and then scanned in STM for detecting the presence or absence of the probe item. The rapid decay of information in STM and the time taken to retrieve it have long been known to be positively related to aging. The question of interest here is whether the Older–Younger RT difference in retrieval time is linearly or exponentially related to the tasks' STM loads. Saul Sternberg's MS paradigm seems ideally suited to examining this question. In a study by Vernon and Vollick (1998), memory load is controlled by using strings of letters that differ in length: either 5, 6, 7, 10, 11, 13, or 14 different randomly selected letters are presented for 2 s on a computer monitor. Following a random interval of between 1 and 2 s, a single probe letter appears on the screen. On a binary response console, the subject lifts his index finger from the home button (RT) and presses either the YES or NO button (MT) to indicate whether the probe letter was or was not a member of the positive set. RT is recorded separately for YES and NO responses. The subjects in the Young group were 70 university students (mean age 19.1 years); there were 70 subjects in the Old group with a mean age of 71.2 years.

As shown in Figure 6.8, both the Young and the Old groups show the typical Sternberg Effect of a linear increase in RT as a function of set size. Figure 6.9 shows a Brinley plot of the same data, viz., the RT means for the positive (YES) responses on each of the seven different string lengths. Although the Old group's RTs are significantly longer than the Young group's (O–Y = 840 ms, on average), the Old RTs are almost perfectly predicted from the Young RTs, with $r_{YO} = .99$. The Brinley plot of the means for the negative RTs present much the same picture, with $r_{YO} = .93$. The negative RTs, which also manifest the Sternberg Effect, are longer than the positive by 65 ms in the Young group and 73 ms in the Old group; the difference of 8 ms is nonsignificant.

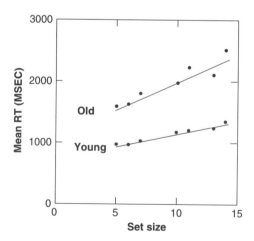

Figure 6.8: Mean RT (for positive responses) as a function of set size (SS = 5, 6, 7, 10, 11, 13, 14) for the Old and Young groups, showing the Sternberg Effect. The Pearson correlations of mean RT with SS in the Old and Young groups are .94 and .98, respectively. (Data taken from Vernon & Vollick, 1998.)

The same Young and Old groups were also given the Posner paradigm, which consists of two measures: (1) the time to discriminate perceptually between a pair of words as being physically the Same or Different (SD task), and (2) the time to discriminate between a pair of words as being Synonyms or Antonyms (SA task), thereby making a demand on retrieval of information from LTM. The difference in RT between these tasks (SA–SD) is considered a measure of the time taken for accessing semantic information in LTM. So we can ask whether the RTs for SA and SD behave in the same manner as RTs in the Sternberg task. Figure 6.10 shows a Brinley plot of the same data as shown in Figure 6.9 but with the mean RTs for the SD and SA tasks included in the plot (indicated by the circled data points). We can see that the mean Same–Different (SD) RT falls on the same line with the seven data points of the STM-scan data, whereas the Synonym–Antonym (SA) task with a demand on retrieval of information in LTM is clearly an outlier, showing a statistically significant departure from the regression line. When the time for the visual discrimination process in the SD task is subtracted from the time for the SA tasks to reveal the time for retrieval of semantic information from LTM, the Younger–Older difference is a nonsignificant 10 ms. The same phenomenon indicating a greater preservation of automatic access to verbal codes in LTM with increasing age than occurs in most other cognitive processes was found in an earlier study by Cerella, DiCara, Williams, and Bowles (1986). Generally, these findings are in accordance with R. B. Cattell's theory of fluid (Gf) and crystallized (Gc) intelligence, whereby Gf declines with age more rapidly and severely than does Gc. Tasks involving STM are known to be more loaded on the Gf factor; whereas LTM is more loaded on Gc, as measured by tests such as vocabulary, general knowledge, and well-learned or automatized mental skills. A study by Bigler et al. (1995) shows evidence for

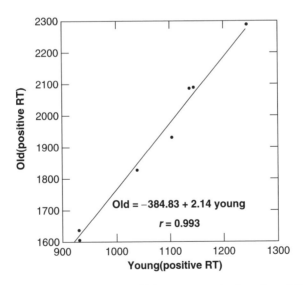

Figure 6.9: A Brinley plot of Young and Old groups on the Sternberg MS paradigm. The data points, from lowest to highest RT, represent SS of 5, 6, 7, 10, 11, 13, 14 randomly selected letters. (Data taken from Vernon & Vollick, 1998.)

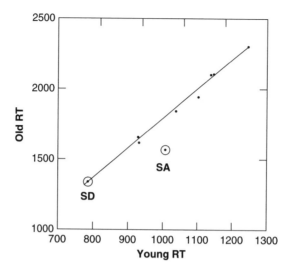

Figure 6.10: A Brinley plot of the Young and Old groups' data as plotted in Figure 6.9, but with the addition of data points (circled) for the mean RTs on the SD and SA paradigms. The one statistically (and theoretically) significant outlier is the data point for SA.

an organic basis for this Gf–Gc difference in relation to age. The decline with age on Gf-loaded tests closely parallels the decrease in brain volume, while the Gc tests do not. Evidently, as illustrated by the observed difference in the effect of aging on the Sternberg and Posner paradigms, there are some mental processes, particularly verbal or lexical acquisitions that are highly consolidated in LTM, that are relatively more spared than others by the global slowing of cognitive functions.

### The Dominance of General Processing Time

The gist of the analyses of age differences in RT across diverse information processing tasks, showing a strong general factor, is not at all unique to the data sets used to illustrate this point. It is found in virtually all sets of published chronometric data that are at all suited to testing the general factor hypothesis. The classic review of this evidence by John Cerella (1985) was based on 189 chronometric tasks from 35 published studies. It showed linear regressions of Old RTs on Young RTs, in which a constant coefficient of proportionality (i.e., the slope of the regression when the intercepts are mathematically constrained to zero) predicts 96 percent of the variance in the older groups' mean RTs across various tasks. The same "slowing coefficients" also predicted the RTs of older subjects from those of younger subjects on 69 more recently published processing tasks. Cerella also concluded from all these data that sensory-motor processes in performing the processing tasks (reflected by the intercept of the regression equation) are less adversely affected by age than are the higher processes of cognition (reflected by the slope of the regression).

The question arises as to whether this general factor of processing time applies only to the divergence of the *mean* RTs between Old and Young groups as task difficulty increases and may not apply to RTs selected from the groups' slower or faster RTs on each task. This question has been clearly answered in a study by Smith, Poon, Hale, and Myserson (1988). They looked at the regressions of Old on Young RTs where the RTs are selected from different percentiles of each subject's total distribution of RTs over trials for each of 20 chronometric task conditions, including various choice RTs, memory scanning, and letter- and word-matching tasks. Errors on these RT tasks occurred on less than 1 percent of the trials and did not differ significantly between the Old and Young groups. The Old *vs.* Young mean RTs at the 10th, 25th, 50th, 75th, and 90th percentiles were examined as Brinley plots, from which the authors commented: "… what holds at the center of the distribution holds all the way through." Figure 6.11 displays this finding as the regression of the Old RTs on the Young RTs for all 20 tasks at each of the five percentile levels (= 100 data points). Considering the diversity of tasks and the levels of RT performance represented in Figure 6.11, the fit of the data points to the predicted values (the upper line) is remarkably good, accounting for 97.6% of the variance in the Old group's means. The authors concluded "old and young subjects' cognitive performances are controlled by the same processes with no added stages or other qualitative differences in performance for the old. Indeed it seems the

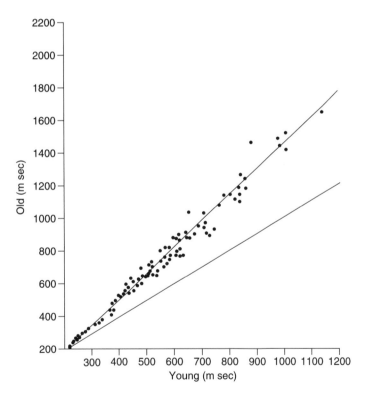

Figure 6.11: Plot of RTs for 20 conditions combined across 5 percentile levels for Old versus Young subjects. The upper line is the fitted regression line; the lower line represents equality for Old and Young RTs. (From Smith et al., 1988; with permission of the *Australian Journal of Psychology.*)

old subjects are merely the young subjects slowed by a multiplying constant, at least on cognitive processes early in practice. That is, aging has a unitary effect on cognitive performance. It is sufficient to postulate a global slowing with age to account for both within-task and between-task aging effects" (Smith & Brewer, 1985, p. 202). And finally, "There is a generalized unitary slowing of cognitive information processing with age, and this slowing similarly affects an individual's best, average and worst performance on easy, moderate and difficult cognitive tasks" (p. 209).

The general processing time hypothesis has been extended beyond the study of aging to group and individual differences of the same age, but that differ in overall RT on a number of diverse processing tasks (Hale & Jansen, 1994). The upper and lower quartiles in this respect (labeled Slow and Fast, each group with $N = 10$) were selected from a group of 40 university undergraduates of similar age. They all were given seven different computerized processing tasks with conditions of varying difficulty, yielding 21 experimental conditions with RTs ranging overall from about 300 to 3000 ms. As seen in Figure 6.12, the mean RTs of the Slow and Fast groups, regressed on the overall mean response latencies (RTs) of the whole sample ($N = 40$), substantiate the general processing time hypothesis as applicable to groups of the same age but differing in ability.

Another question that stems from this study is whether the regularity shown by the general processing time phenomenon holds for individual differences as well as for mean differences between groups. Hale and Jansen (1994) also examined this question with their data by plotting the RTs for individual subjects as a function of the overall mean group RTs on the same 21 processing tasks used in the analysis shown below (Figure 6.12). They found the

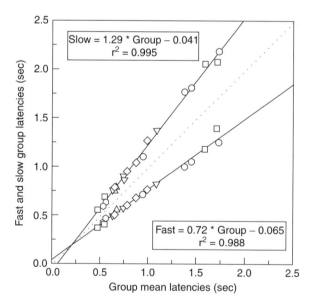

Figure 6.12: Fast and Slow groups' RTs as a function of total groups mean RTs. (Reprinted from Hale & Jansen, 1994, with permission of the American Psychological Society). *Note*: The asterisk (*) in the regression equations stands for multiplication.

same linear regressions for individual subjects selected from the high, middle, and low segments of the total distribution of individual RTs. Linear regression predicted individuals' RTs for the 21 tasks remarkably well from the overall group means, showing correlations between individual RT scores and group mean RTs ranging from .87 to .98, with a mean of .96. Myerson et al. (2003) have shown that the GPT model that applies to differences between healthy younger and older adults and to differences between healthy individuals who differ within the normal range of psychometric *g* (or IQ). The GPT model also holds for RT comparisons between healthy adults and people with certain health problems such as multiple sclerosis, and for those who are mentally retarded, depressed, schizophrenic, or brain injured. But the topic of RT in pathological conditions is reviewed in Chapter 13.

## Micro-Effects of Cognitive Aging on Response Times

It is now well established that a large global factor accounts for most of the variance in the effects of aging in normally healthy individuals across a wide range of various response-time measurements. Moreover, it appears that the same, or a highly similar, general factor accounts for most of the variance in response-time measurements during the period of development from early childhood to maturity, and this factor is closely related to Spearman's *g* factor derived from diverse psychometric tests. But the presence of such a broad general factor in RTs should not override and discourage examination of other reliable but "fine grained" effects of aging. Coarser or finer scales of measurement or of statistical analysis are appropriate for different RT phenomena. In the Brinley plots previously shown, for example, certain tasks produce deviations from the regression line predicted by the global factor amounting to 70 ms or more, which, if proved statistically reliable and replicable, should warrant further analysis. But the "fine-grain" of RT actually involves much more than just the reliable deviations of certain tasks from the general regression line in Brinley plots.

The three main sources of reliable RT variance independent of the general factor are: (1) cognitive processing tasks in which performance depends on different neural systems, such as perceptual discrimination and decision speed versus the speed of memory and retrieval processes; (2) certain pathological conditions that selectively affect only certain information processes; and (3) measurements of phenomena that constitute a different, more close-up, microscopic, or fine-grained level of analysis of RT besides an individual's mean or median RT *per se*, such as intraindividual trial-to-trial variation (e.g., as a measure of "neural noise") that is independent of mean RT, sequential effects on RT across trials following response errors, and changes in strategies such as the speed–accuracy trade-off. Topics 1 and 2 are taken up in later chapters. A few prominent examples of 3 are briefly summarized here.

### *Speed–Accuracy Preference in Relation to Age*

It is commonly believed that impulsiveness decreases and cautiousness increases with age. If so, this implies the possibility that the slowing of RT in cognitive tasks is due not to an increasing general capacity limitation in speed of information processing, but results from a shifting speed–accuracy trade-off function with increasing age. Assuming speed–accuracy trade-off also applies to performance on conventional intelligence tests, it also implies that this may be the common factor accounting for at least some part of the negative correlation of age with raw scores on IQ tests and with speed of processing in elementary cognitive tasks. Such a finding would raise the more fundamental question of whether the measured age declines in cognitive abilities are actually *construct relevant* to the hypothesis of a general slowing of processing rate with age or is only *performance specific*, depending on the interaction of personality traits (e.g., impulsiveness, extroversion) with specific task conditions that allow the expression of these traits via speed–accuracy trade-off preferences. It should be noted that such preferences may not be based on conscious or willful decisions but may be an unconscious aspect of an individual's personality and may also unknowingly change with age.

A study by Phillips and Rabbitt (1995), though not definitive on all aspects of this question, throws some light on it. In a quite ideally selected sample of older adults, aged 50–79, impulsiveness in test-taking was measured in a battery of conventional verbal and nonverbal paper-and-pencil tests, each administered with a time limit. On each test, *accuracy* was measured as the percentage of correct answers on the attempted items. *Speed* was measured as the number of items attempted within the time limit. *Impulsivity* was measured as the difference between standardized scores on speed and accuracy, i.e., Impulsivity = ($z$ speed $-$ $z$ accuracy). This measure proved to be a quite substantial general factor common to all four of the conventional tests, qualifying the Impulsivity measure as an authentic personality factor that affects test-taking behavior. (The mean correlation among the four measures of Impulsivity was .49 and their PC1 accounts for 62 percent of the variance among the Impulsivity measures on the four tests, indicating a remarkably large common factor.) Surprisingly, however, this impulsivity factor is largely independent of total correct answers on the tests (average $r = -.03$). The Impulsivity score showed no significant correlations with the trait of extroversion (Eysenck Personality Inventory), in which impulsivity has been claimed to be a component. The Impulsivity measure also showed no significant correlation with a questionnaire assessment of impulsivity. The Impulsivity scores estimated from conventional tests, however, showed very small but occasionally significant negative correlations (average $r = -0.17$) with the RT measures on the three elementary information processing tasks. (These processing tasks themselves had error rates too small to allow reliable estimation of impulsivity scores.) The authors concluded, "None of the relations between intelligence test impulsivity scores and age were significant. It therefore seems unlikely that age deficits in intelligence test performance can be attributed to speed-accuracy preferences changing with age" (p. 24).

A study by Smith and Brewer (1995), amplifying the results of a previous study (Smith & Brewer, 1985), takes an even more microscopic view of age difference in RT. It focused on age differences associated with *response errors* in a 4-choice RT task. The Young group was aged 18–30; the Old group 58–75 years. Except for the overall mean difference between these groups, the shapes of the distributions of RTs were almost identical across both groups, as attested by the fact that at any point in the distribution a linear regression equation almost perfectly predicts Old RT from Young RT (i.e., Old RT = 14 + 1.12 × Young RT, with $r^2 = .99$). Thus an overwhelming unitary factor is operative on the mean

RTs in both groups. The overall error rate on this RT task was about 4 percent, with a non-significant Old–Young difference of 1.3 percent.

As is typical of other studies, for both the Old and Young groups, error responses consistently had longer RTs than do correct responses. Yet, despite the great similarity between age groups in all these features of RT performance, Smith and Brewer managed to find a difference in which the older group's performance appears to differ from the younger group's performance. This in no way contradicts the hypothesis of a general slowing factor with age, but may be regarded as one of its micro-manifestations. The phenomenon is, in fact, named "micro-caution" by the authors. Both correct and error responses are slower for the older group. Older adults, on average, also show less fine control of RTs and take fewer risks by adopting a less efficient strategy of keeping their responses in a more "safe" bandwidth of RTs, but they *over*-adjust, making slower than their average RTs, in the trials immediately following a response error. Like younger adults on error trials, older subjects tend to respond faster and faster on successive trials until they make an error, after which they suddenly slow down and begin the risk-taking cycle over again. Individual differences in this respect are considerable, however, and as Smith and Brewer (1985) have noted, the performance of some older and younger subjects are indistinguishable in these features of RT performance that show up so clearly in the group means. They state "Overall, we are left with the conclusion that the older group used the same processing mechanisms as the younger group in this task. A variety of indicators suggests that older participants respond more carefully or do not push themselves to their limits as often as do younger participants. This is underpinned by coarser RT adjustment, and the result is less efficient RT constraint within the target — fast, safe — RT bands, and hence an increased variance and overall mean RT" (Smith & Brewer, 1995, p. 246). These micro-effects of RTs preceding and following an error RT are shown in Figure 6.13.

But the most important aspect of this finding is that the virtually identical effect is found in comparisons of children with young adults, as shown in a study by Brewer and Smith (1989) previously described in Chapter 5 (see Figure 5.11). The older adults in Figure 6.13 most closely resemble the analogous graph for thirteen-year-olds in Figure 5.11. In other words, if the RT performance of older persons is compared with that of people who are very much younger and still in the developmental period — late childhood or early teens — the similarity of Old and Young RT performance, even at the micro-level of analysis, is clearly evident. Strangely, this striking regularity across the wide age gap is not mentioned in the later paper by Smith and Brewer (1995), nor is their earlier 1989 paper even cited. Yet the authors offer much the same explanation for the findings of both studies. It appears most likely that the observed effects of regulation of RTs in accord with error responses is really part and parcel, rather than a fundamental cause, of the development and decline of the general factor that dominates RT phenomena across the entire age range in normal, healthy individuals.

*Differentiation and Dedifferentiation Hypotheses.* The *differentiation hypothesis* holds that cognitive ability becomes increasingly *differentiated* throughout the developmental period. In terms of factor analysis, the common factor variance among diverse mental tests becomes increasingly differentiated from a general factor into a number of group factors and specificity.

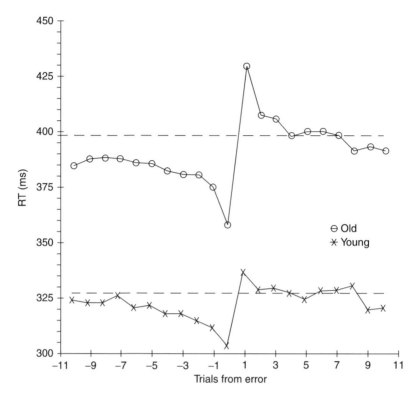

Figure 6.13: Old and Young groups' RTs on 10 correct trials preceding and 10 trials following an error response. The dashed horizontal lines represent the mean correct RTs for each group. (Reprinted from Smith & Brewer, 1995, p. 244; with permission of the American Psychological Association.)

In later maturity, the opposite trend occurs, termed *dedifferentiation*, as the proportion of common factor variance gradually increases while there is a shrinkage of group factors and specifics. Some of the chief studies of these effects are cited elsewhere (Carroll, 1993, pp. 677–681; Deary et al., 1996; Hertzog & Bleckley, 2001; Li & Lindenberger, 1999). The effects of differentiation and dedifferentiation are generally not great, as a large general factor dominates nearly every test correlation matrix at every age. Evidence for differentiation effects is often equivocal in psychometric test batteries; the effects are relatively small, the factor structure *per se* (i.e., the number and identity of factors) of multi-test batteries usually remains the same across the whole testable age-range, while the proportions of total variance accounted for by each of the factors typically shift slightly in the direction consistent with differentiation or dedifferentiation, depending on the age of the subjects. These effects are presently regarded as methodologically problematic in psychometrics, more so as regards differentiation in the ontogenesis of cognitive abilities than with dedifferentiation in later maturity and old age, which is quite well substantiated. It is likely that further investigations using chronometric methods, because of their absolute scale property, will provide more definitive analyses of the differentiation and dedifferentiation phenomena than has been possible with psychometric tests. Because the investigation of these phenomena involves both the correlations among tests at different ages along with changes in the variance of test scores, factor analysis based on the variance–covariance matrix, rather than on the correlation matrix, seems more appropriate. (Correlation coefficients are standardized covariances, in effect equalizing the total variances across all tests at all age levels.) Correlation coefficients are justified in the factor analysis of various psychometric tests because their scaled scores seldom assure equal metric units, whereas chronometric measures, like the measurement of age, have this analytically valuable scale property, thereby making the factor analysis of variance–covariance matrixes an authentically meaningful procedure. Such investigation seems worth pursuing in light of the fact that these behavioral cognitive phenomena are paralleled by physiological variables in studies of brain imaging and measures of cerebral blood flow (studies cited in Li & Lindenberger, 1999).

Most of this research to date has used conventional psychometric tests, usually given under timed or speeded conditions. An impressively large study by Hertzog and Bleckly (2001) has shown the predicted effects of dedifferentiation on first-order factor loadings in a battery of 25 psychometric tests representing seven primary ability factors. In addition to taking the actual tests themselves, subjects were later also provided with correct answers for each test, which they simply had to copy on the appropriate answer sheets as a speeded test to create an "Answer Sheet" factor. Some ability tests were more highly correlated with the Answer Sheet factor than were others and these correlations varied with age in a sample of 622 adults ranging from 42 to 78 years. The Answer Sheet speed factor, which presumably involved more sensory-motor than cognitive ability, accounted for some but not all of the dedifferentiation effect in the regular tests that used the same kinds of answer sheets. Because the sheer "mechanics" of the answer sheets *per se* have differential effects related to age, tests measuring the same factor show age differences in test scores that are not construct-intrinsic to the latent ability factors. Noting the crucial distinction between age changes in performance-specific versus construct-relevant processing speed, the authors concluded, "Findings were consistent with the view that speed of information processing can be both an important correlate of individual differences in rates of intellectual aging and a performance-specific confound that distorts estimates of age-related changes in psychometric ability tests" (Hertzog & Bleckley, 2001, p. 191). Such distorting confounds can be more readily controlled in chronometric studies in which the same apparatus is used for all tests and the tests can be so graded in difficulty as to establish for every subject reliable baseline control data for the sensory-motor times for the mechanical aspects of performance. For example, simple RT can serve as a baseline or covariate control in measuring individual differences in various degrees of choice or discrimination RT (Jensen & Reed, 1990).

### *Percentiles of the RT Distribution in Relation to Age*

The unit of analysis in most studies of the effects of aging on RT is the subject's mean or median RT over a given number of trials. Another micro-effect of RT in relation to age concerns whether age differences in RT are constant throughout the full range of RTs produced

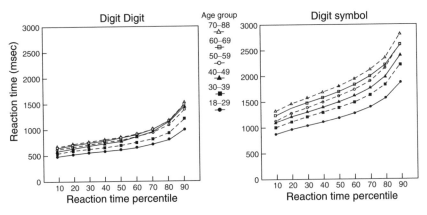

Figure 6.14: Mean reaction times on two processing tasks plotted as a function of their percentiles in the total distribution of 180 test trials for six age groups. (From Salthouse, 1998, with permission of Elsevier.)

by subjects on the same task in a given number of trials. Each individual's RTs vary in length from trial-to-trial, from very short to quite long. Salthouse (1998) investigated this question by plotting the mean RTs in 180 trials from different RT percentiles for each individual (in nine steps from 10th to 90th percentile) for each of six age groups ranging from 18 to 88 years. The different age groups, although predictably differing in overall RT, were alike in showing the same pattern of RTs plotted as a function of their percentiles in the RT distribution. There is an almost constant difference between age groups in all segments of the distribution of RTs, as shown in Figure 6.14. The correlations between RT and age, however, were not uniform across all percentiles; in five different processing tasks the highest age × RT correlations occurred in the low to middle range of percentiles (10–50th), with correlations between .4 and .6. The somewhat lower correlations for the slowest RTs (80–90th percentiles) are associated with the greatly increased variance and probably lower reliability of the slowest RTs in the total distribution. A similar picture was found for correlations between psychometric cognitive ability scores and RTs at different percentile levels. RTs from different percentiles of the RT distribution appear qualitatively the same with respect to age differences across nearly the entire range of individual RT distributions.

### Intraindividual Variability as a Function of Age

Measured as an individual's *SD* of RT across a given number (*n*) of test trials, intraindividual variability in RT (labeled RTSD) is much less well understood at present than is warranted by its possible importance, not only for the study of normal cognitive aging, but also for the study of pathological organic brain conditions. Its relevance to aging is betokened by the *neural noise* hypothesis of cognitive aging. Neural noise is a hypothetical construct proposed to account for the trial-to-trial variation in an individual's RT. Its increasing with age has been explained hypothetically as an increase in the signal–noise ratio associated with the fidelity in neural transmission of information in the brain, probably at the synaptic junctions. It is a reasonable hypothesis, especially in view of recent evidence at the biological level, such as the age-related variability in latency of the P300 wave of the average evoked potential, which is also correlated with other cognitive measures of information processing such as the latency of stimulus apprehension and evaluation and even IQ in some studies (Li & Lindenberger, 1999). What is most well established is that RTSD, like mean RT, increases with age beyond the early years of maturity. In fact, RTSD follows the same trajectory as RT across the whole life span, from early childhood to old age. As a result of such observations some theorists have suggested that both RT and RTSD are merely two different measures of one and the same latent variable. Though the RT and RTSD measures are phenomenally different, like the diameter and circumference of a circle, they could both reflect one and the same latent process and therefore be as perfectly correlated as measurement error would permit. The present data on this point, however, seem quite inconclusive, because it has also been found that RT and RTSD have significantly different correlations with other general cognitive measures, such as IQ and psychometric *g*, even when equated for reliability. (RTSD generally has the higher correlation [Jensen, 1992].) In some of these studies, however, the results are possibly contaminated by an artifact of using the individual's *median* RT, rather than the *mean* RT, in comparing or correlating RT with RTSD. The high but less than perfect correlation between median and mean RT could possibly account for the apparent differences in the correlations of RT and RTSD with other cognitive variables.

If RT and RTSD have only one and the same factor in common, the loadings of this factor on different processing tests should be virtually the same or highly similar. In Ananda's (1985) study of nine information processing tasks taken by 76 elderly people (ages 51 to 87 years), the PC1 was extracted separately from two different measures derived from the correlations among the same set of tests: (1) the *median* RT and (2) the RTSD. The column vectors of the nine factor loadings for the RT and RTSD factors are correlated only .044. Ananda also correlated both median RT and RTSD for each of the nine tests with age. The average correlations of RT and RTSD with age (in the range 51 to 87 years) were .273 for RT and .154 for RTSD. But the more important point is that the column vector of nine correlations between RT and age and the corresponding column vector of correlations between RTSD and age scarcely resemble each other; they are correlated only .38. Across the different processing tests there was also little consistency in the patterns of RT and RTSD correlations with measures of intelligence (Raven Matrices and Vocabulary). These considerable inconsistencies between RT and RTSD are too great to be explained by differences in reliability or by use of the median rather than the mean RT. They assuredly underline the question of whether RT and RTSD both reflect entirely one and the same latent variable or somewhat different latent variables.

The reported correlations between RT and RTSD are rendered questionable when both measures are from the same set of data, hence are not experimentally independent; that is, there is an inevitable correlation between means and SDs when both are derived from the same set of RT measurements. In a methodologically correct study of the median RT × RTSD correlation in which median RT and RTSD were measured on the Hick paradigm on two occasions, ensuring that they were experimentally independent. Individuals' median RT and RTSD were correlated .67 after correction for attenuation, using the test–retest reliability coefficients of each variable obtained from the same data sample of 50 university students (Paul, 1984, reported in Jensen, 1987a). Unfortunately, only median RT was measured in this study, although it seems doubtful that the use of mean RT would have raised the RT × RTSD correlation to approximate unity.

Another method for examining the RT–RTSD relationship is by the coefficient of variation (CV = SD/Mean), which, for each individual, measures the proportionality between RTSD and RT. The CV is a meaningful statistic only when both variables are derived from ratio scales, as are chronometric measures, unlike nearly all psychometric measures. If the CV for a given processing task remains constant across individuals within narrow limits of measurement error, it signifies that mean RT and RTSD are simply functions of the same latent variable and hence are redundant measures of individual differences in that variable. I have not come across any study that adequately tests this hypothetical outcome. However, it is consistent with the hypothesis that individual differences in RT and in RTSD are multiplicative functions of some single common factor. A rigorous test of this hypothesis seems an essential undertaking for the investigation of the "neural noise" theory of cognitive aging.

Each of the individual's RTs obtained in a given period of time can be thought of as a random sample from a theoretical distribution of possible RTs characteristic of that individual. Each parameter of the distribution, such as the first four moments and other derived indexes, potentially constitutes an individual difference variable. An individual's mean RT and RTSD over $n$ number of test trials, for example, are conceived as statistical estimates of the mean and SD of the theoretical distribution of all the potential RTs characteristic of the individual in a particular test session. This model can be elaborated for two or more mixed or superimposed distributions that are the hypothesized underlying sources of the individual's RT variation. These could be, for instance, an amalgam of the distributions for the sensory-motor component and the cognitive component of RT.

A key question essential to this line of inquiry is whether variation in an individual's trial-to-trial represents random variation in long runs of error-free trials. Such randomness is implied by the neural noise theory of cognitive aging. But there are better measures of "noise," randomness, or entropy in an individual's RTs over $n$ successive trials than the SD. The mean square successive difference (MSSD) between successive RTs (see Chapter 4, p. 67) is probably the best example. Reliable downward or upward trends in RT across trials due to a practice effect or to fatigue are reflected by the RTSD along with more purely random variation. The MSSD is independent of such systematic trends across trials. Randomness implies independence and unpredictability, such that within the numerical boundaries of an individual's succession of RTs, the RT on any given trial cannot be predicted better than chance from the RT on any previous trial. The proviso of specifying error-free trials is required because we know that the RTs on a few trials preceding and following an error response have atypically shorter or longer RTs compared to the RTs in the runs of trials between error responses and are therefore somewhat predictable by the occurrence of an error. So one should measure RT "noise" in runs of error-free trials at some distance (say ±5 trials) from any error trial. Intraindividual variability in error-free RTs are more likely to detect trial-to-trial randomness than measures indiscriminately based on all of the test trials. As simple RT is virtually error-free, it would be a good paradigm for the study of whether variation is random in individuals' RTs across trials, by applying statistical tests of randomness, such as the autocorrelation between successive trials and the Von Neumann Ratio (i.e., for a given individual, the MSSDs in RT across trials divided by the variance of all the RTs; see Chapter 4, p. 67). Significance tests are available using the percentiles of the Von Neumann Ratio for random numbers, which have been tabulated by Hart (1942). The measurement of "neural noise" and its possible role in cognitive aging indeed seems virtually virgin territory for chronometric exploration.

# Chapter 7

# The Heritability of Chronometric Variables

Evidence for a genetic component in chronometric performance assures us that reliable measures of individual differences reflect a biological basis and that variation in the brain mechanisms involved are not solely the result of influences occurring after conception. Any nongenetic effects could be attributable to exogenous biological influences, such as disease or trauma, or to nutrition in early development, or to purely experiential effects that transfer to an individual's performance on chronometric tasks, for instance, practice in playing video games.

In addition to heritability of individual differences in performance on a given chronometric task, it is important to know the nature of the task's correlation with other variables that lend some degree of ecological validity (i.e., the variable's correlation with 'real life' performances generally deemed as important in a given society). Two distinct variables, for example, reaction time (RT) and IQ, could each be highly heritable, but the correlation between them could be entirely due to nongenetic factors. The two variables, say, RT and IQ, could each be indirectly correlated with each other via each one's correlation with a quite different variable that causally affects both variables, for example, some nutritional factor. On the other hand, two variables could be genetically correlated. A genetic correlation, which is determined by a particular type of genetic analysis, indicates that the variables have certain genetic influences in common, though other specific genetic or environmental factors may also affect each variable independently. In the terminology of behavioral genetics, both genetic and environmental effects may be either *common* (or *shared*), in whole or in part, between two or more individuals, or they may be *specific* (or *unshared*) for each individual.

There are two types of genetic correlation between kinships: (1) *simple* genetic correlation and (2) *pleiotropic* correlation.

In a *simple genetic correlation* different genetic factors, a and b, for different phenotypic traits, A and B, are correlated in the population because of population heterogeneity due to cross-assortative mating for the two traits. Hence within any given family, in meiosis, the genes for each trait are independently and randomly passed on to each of the offspring. Because of independent, random assortment of the genes going to each sibling, the causal connections a→A and b→B themselves have no causal connection in common. A well-established example is the population correlation between height and IQ. These phenotypes are correlated about .20 in the general population, although there is no causal connection whatsoever between genes for height and genes for IQ, as shown by the fact that there is zero correlation between the height and IQ of full siblings (Jensen & Sinha, 1993). All of the population correlation between height and IQ exists *between* families; none of the correlation exists *within* families (i.e., between full siblings).

In a *pleiotropic correlation*, a single gene has two (or more) different phenotypic effects, which therefore are necessarily correlated *within* families. The sibling who has the pleiotropic gene will show *both* phenotypic traits; the child who does not have the gene will show neither of the traits. An example is the double-recessive gene for phenylketonuria (PKU), which results in two effects:(1) mental retardation and (2) a lighter pigmentation of hair and skin color than the characteristic of the siblings without the PKU gene. (Nowadays, the unfortunate developmental consequences of PKU are ameliorated by dietary means.) Another likely example of pleiotropy is the well-established correlation (about +.25) between myopia and IQ. The *absence* of a within-family correlation between two distinct phenotypic traits contraindicates a pleiotropic correlation. The *presence* of a within-family correlation between two distinct phenotypes, however, is not by itself definitive evidence of pleiotropy, because the correlation could possibly be caused by some environmental factor that affects both phenotypes. Pleiotropy can be indicated by the method of cross-twin correlations between two distinct variables (e.g., different test scores), A and B. The method is applicable to both monozygotic, twins reared apart (MZA) and monozygotic, twins reared together –dizygotic, twins reared together (MZT–DZT) designs. Pleiotropy is indicated if the MZA twins in each pair (labeled MZA1 and MZA2), show essentially the same cross-correlations on tests A and B, i.e., the cross-correlation between test A scores of MZA1 and test B scores of MZA2, and twins 2 are significant and virtually the same as the correlations between A scores of MZA1 and B scores of MZA1 (and also the same for MZA2). The two main types of genetic correlation, simple, and pleiotropic, are illustrated in Figure 7.1, the direct and cross-twin correlations in Figure 7.2.

## Heritability Defined

Heritability, symbolized as $h^2$, is a statistic derived from various kinship correlations on a given metrical trait that estimates the proportion of the total phenotypic variance in a defined population that is attributable to genetic variation, i.e., individual differences in genotypes. (An exceptionally thorough critical exposition of the meaning of heritability is provided by Sesardic, 2005.)

In the simplest formulation, a phenotype ($P$) is represented as the sum of its genetic ($G$) and environmental ($E$) components, i.e., $P = G + E$. The phenotypic variance ($V_p$) therefore,

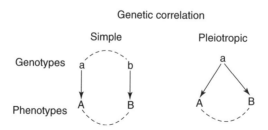

Figure 7.1: Genetic correlations, simple and pleiotropic, between phenotypes A and B. Arrows indicate a causal relationship between genes and phenotypes; dashed curved lines indicate a statistical correlation.

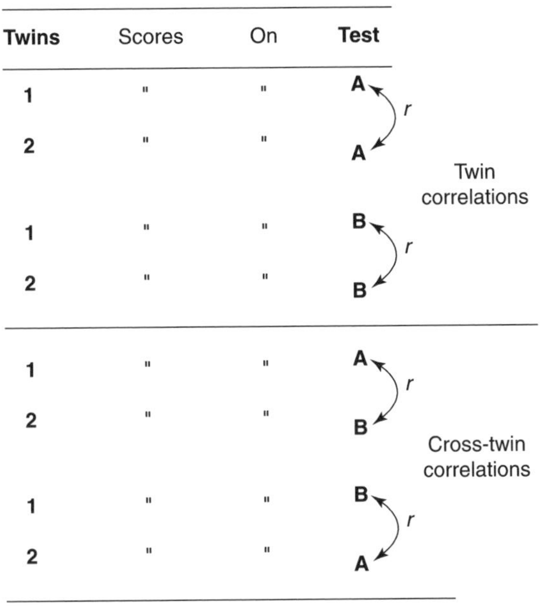

| Twins | Scores | On | Test |
|-------|--------|-----|------|
| 1 | " | " | **A** |
| 2 | " | " | **A** |

)*r*

Twin correlations

| 1 | " | " | **B** |
| 2 | " | " | **B** |

)*r*

| 1 | " | " | **A** |
| 2 | " | " | **B** |

)*r*

Cross-twin correlations

| 1 | " | " | **B** |
| 2 | " | " | **A** |

)*r*

Figure 7.2: Illustration of twin and cross-twin correlations, which are usually obtained from samples of both MZ and DZ twins and are used in quantitative genetics to estimate the genetic correlation (or the environmental correlation) between different metric traits. For further discussion and references explicating this method, see Plomin, DeFries, and McClearn, 1990, pp. 235–246.

is $V_P = V_G + V_E$. Heritability, then, is defined as $h^2 = V_G/V_P$. The proportion of phenotypic variance attributable to the environment ($e^2$), i.e., variance due to all nongenetic factors, is not measured directly, but is the residual: $e^2 = 1-h^2$.

The two most commonly used kinds of data for estimating $h^2$ in humans are based on (1) the correlation between identical, or MZA; and (2) contrasting the correlation between MZT with the correlation between fraternal, or DZT. *Note*: the effect of age differences between the twin-pairs in such studies are statistically removed.

Here is the logic in estimating $h^2$ from these twin correlations:

(1) MZA have all of their genes in common (i.e., identical genotypes) but they have no shared environment; so, the intraclass correlation between MZA for a given trait is entirely a result of they having all their genes in common. Therefore, the intraclass correlation between MZA twins, labeled $r_{MZA}$, is a direct estimate of the proportion of genetic variance in the measured trait, so $r_{MZA} = h^2$.[1]

(2) The correlation between MZT, $r_{MZT}$, comprises all of their genes and their shared environment, and the correlation between DZT, $r_{DZT}$, comprises half of their genes and their shared environment. So the difference $r_{MZT}-r_{DZT}$ estimates half of the genetic variance; therefore $h^2 = 2(r_{MZT}-r_{DZT})$.

## Estimates of $h^2$ in Chronometric Variables

The total published empirical literature on this subject consists of 12 independent studies, as far as I can determine. It is difficult to summarize or compare them all in detail, as they are so heterogeneous in the various chronometric paradigms, procedures, specific measures, and subject samples that they used. For this specific information readers are referred to the original sources, which for economy are cited here only by first author and date of publication: Baker, Vernon & Ho (1991), Boomsma (1991), Bouchard et al., (1986), Couvée (1988a,b), Ho (1988), Luciano et al., (2001), McGue et al., (1984, 1989), Neubauer (2000), Petrill (1995), Rijsdijk (1998), and Vernon (1989). Three of the studies (Bouchard, 1986; McGue, 1984, 1989) are based on MZA; all the others are based on MZT–DZT. Both of these types of studies provide fairly similar estimates of $h^2$, although those based on MZA are less variable, generally they give slightly higher values.

The findings of these studies fall into three categories

(1) the $h^2$ of *direct* measures of chronometric variables (RT or speed of information processing);
(2) the $h^2$ of *derived* measures (*intraindividual variability* in RT (RTSD), *difference scores* (such as name identity–physical identity (NI–PI) in the Posner paradigm), and the *intercept* and *slope* of RT (as in the S. Sternberg memory scan paradigm) as a function of information load or task complexity; and
(3) the *genetic correlation* ($r_{GP}$) between a direct chronometric variable (e.g., RT) and a psychometric variable (e.g., IQ or g). A genetic correlation of this type is pleiotropic (see Figure 7.1).

The results of these studies can be summarized accordingly:

(i) For direct measures of RT (speed of response), the estimates of $h^2$ are Mean = .44, Median = .48, and SD = .19.

(ii) For derived measures (RTSD, NI–PI difference scores, intercept, slope in S. Sternberg paradigm), $h^2$ Mean = .29, Median = .23, and SD = .21. Omitting RTSD, the values of $h^2$ based only on difference scores, intercept, and slope are Mean = .20, Median = .20, and SD = .08.

(iii) For genetic correlations between direct measures of RT and IQ, Mean = .90, Median = .95, and SD = .134. In other words, there is a very high degree of genetic determination of the correlations between RT and IQ. The phenotypic correlation between RT and IQ is lowered by nongenetic effects.

Authors of the various studies concluded the following:

"The common factor (among 11 RT tests) was influenced primarily by additive genetic effects, such that the observed relationships among speed and IQ measures are mediated entirely by hereditary factors" (Baker, 1991, p. 351).

"…the general mental speed component underlying most of these tasks, which are strongly related to psychometric measures of *g*, do show substantial genetic effects" (Bouchard, Lykken, Segal & Wilcox, 1986, p. 306).

"The phenotypic relationship between the measures of general intelligence and the measures of speed of processing employed are due largely to correlated genetic effects" (Ho, Baker & Decker, 1988, p. 247).

"The results reported here support the existence of a general speed component under-lying performance on most experimental cognitive tasks which is strongly related to psychometric measures of *g*, and for which there are substantial genetic effects" (McGue, Bouchard, Lykken & Feier, 1984, p. 256).

"The RT tests' correlations with full-scale IQ scores correlated .603 with their heritabilities, and the tests' correlations with *g*-factor scores extracted from the 10 subtests of the Multidimensional Aptitude Battery correlated .604 with their heritabilities" (Vernon, 1989b).

A large study of the genetic correlation between speed of information-processing (SIP) and IQ in MZT and DZT twins tested at ages between 16 and 18 years concluded, "Multivariate genetic analyses at both ages showed that the RT–IQ correlations were explained by genetic influences. These results are in agreement with earlier findings ... and support the existence of a common, heritable biological basis underlying the SIP–IQ relationship" (Rijsdijk, Vernon & Boomsma, 1998, p. 77).

Another large study of MZT and DZT twins concluded, "To summarize, SIP — as measured by ECTs — has been shown to correlate substantially with psychometric intelligence. Variance in ECTs shows substantial genetic influence, although somewhat less than psychometric intelligence. The phenotypic correlations between measures of mental speed and psychometric intelligence are largely, but not exclusively, due to overlapping genetic effects. The same can be concluded for the correlations among independent measures of intelligence. Thus, the speed of information processing is partially heritable and shares part of its genetic influence with psychometric intelligence. But there are also specific genetic influences apart from those pleiotropic genes, affecting both psychometric intelligence and mental speed" (Neubauer, Spinath, Riemann, Borkenau & Angleitner, 2000, p. 285).

In order to obtain the possibly truest overall estimate of $h^2$ based on all 12 of he extant studies, Beaujean (2005) undertook a sophisticated model-fitting meta-analysis of all the reported data, a type of analysis that, in effect, partials out the irrelevant variance resulting from the marked heterogeneity of the elementary cognitive tasks (ECTs), the different types of twin studies, and the diverse subject samples used in the 12 extant data sets. His "best estimate" of the average $h^2$ of the various RT measures is .53, which is hardly different from the average $h^2$ of various psychometric tests obtained under similarly heterogeneous conditions. The subjectively simpler ECTs averaged lower heritability (.40) than the subjectively more complex ECTs (.67).

### Generalizations from the Present Studies of the Heritability of Chronometric Variables

(1) Direct measures of RT or speed of information processing are unquestionably heritable, with a wide range of $h^2$ values (depending on various conditions) that are generally somewhat smaller than the values of $h^2$ for most psychometric tests, although there is considerable overlap between the distribution of $h^2$ values for direct chronometric variables (RT) and the distribution of $h^2$ values for psychometric test scores. The lower values of $h^2$ for RT than for test scores seems to be related to the relative

simplicity and homogeneity of most RT measures and to their greater specificity (i.e., sources of variance unique to each different RT task).

(2) There are two methods of data analysis that invariably increase the estimated $h^2$ of RT measures: (a) a *composite* RT, and (b) a *latent trait* analysis of RT.

   (a) The *composite* RT is the overall average RT (either mean or median) obtained from an individuals' mean or median RTs derived from a number of *different* ECTs. The composite mean RTs can be either the raw RTs or they can be the means of the standardized RTs within each ECT. The $h^2$ based on the raw composite RTs is generally higher than the $h^2$ based on standardized RTs. The reason is that raw RTs of different ECTs reflect their different levels of complexity or difficulty, and the more difficult tasks have larger values of RT. The larger values of RT (up to a point) also have higher $h^2$, and therefore the tasks composing the raw composite are, in effect, weighted in favor of yielding a higher value for $h^2$ in the total composite than does the standardized RT. The raw composite RT takes advantage of what has been termed the "heritability paradox" (explained in the section on "The 'Heritability Paradox' "). The raw composite value is also preferable because it retains the RT values in the original and meaningful ratio scale of time measurement. The aggregation of different RT measurements in a composite increases $h^2$ simply because it gives increased relative weight to the common factor of RT among a number of different tests, and thereby increases the reliability of the common factor and diminishes the proportion of variance attributable to uncorrelated factors among different RT tasks, that is, their *specificity*. Nevertheless, task specificity remains as a source of variance that attenuates the true correlation between the phenotypic common factor of the various RTs and other variables of interest, including its genotype. (Recall that $h^2$ is the square of the phenotype X genotype correlation (see Note 1).

   (b) Probably the best available method for estimating the $h^2$ of RT is by means of a latent trait analysis or a factor analysis, where the latent trait or factor is the largest common factor of a number of different RT measures of the particular ECTs of interest. This is best represented by the first principal factor (PC1) in a common factor analysis. The resulting factor score, which best represents the common factor among the ECTs, then, is a weighted composite score comprising all of the ECTs in which the RTs on each ECT are weighted by their loadings on the PF1. As is true for all factor scores, there remains some small degree of error consisting of other minor factors, test specificity, and unreliability. These unwanted sources of individual differences variance are never perfectly eliminated, but are more minimized by the use of factor scores (particularly when obtained by the method first proposed by Bartlett, 1937) than by perhaps any other present method.

### *Age Correction*

Age needs to be taken into account in the estimation of $h^2$, because $h^2$ is known to vary across the total age range, generally increasing gradually from early childhood to later maturity. RT, too, varies with age, as was shown in Chapters 4 and 5. As most of the studies of RT heritability are based on groups that are fairly homogeneous in age (mostly young adults), the age factor is not an importantly distorting source of variance. In groups that are more heterogeneous in age, the age factor can be statistically removed from the RT data by means of multiple regression, where the regression equation for each subject includes chronological age (in months), age squared, and age cubed as the independent variables, with RT as the dependent variable; the differences between the obtained and predicted RTs are the age-corrected scores used for estimating $h^2$. This regression procedure is typically sufficient to remove all significant linear and nonlinear effects of age from the RT measures.

### The "Heritability Paradox"

This is the name given to the frequent and seemingly surprising finding that $h^2$ *increases* as a function of task complexity and also as a function of the degree to which tasks call for prior learned knowledge and skills. The notion that this is paradoxical results from the expectation of some theorists that the more elemental or basic components of performance in the causal hierarchy of cognition, going from stimulus (the problem) to response (the answer), are far less removed from the underlying brain mechanisms than are the final outcomes of complex problem solving. The tests of greater complexity are considered to be more differentially influenced by environmental conditions, such as prior cultural–educational opportunities for having acquired the specific knowledge or skills called for by the test. Therefore, it is expected that individual differences in tests that are more dependent on past-acquired knowledge and skills, such as typical tests of IQ and scholastic achievement, should reflect genetic factors to a lesser degree than RT on relatively simple ECTs, which are assumed to depend almost exclusively on the most basic or elemental cognitive processes. Presumably, individual differences in RT on such relatively simple ECTs scarcely involve differences in environmental factors influencing specific cultural or scholastic knowledge and skills, and therefore ECTs should have higher $h^2$ than the more experientially loaded psychometric tests.

It turns out, however, that the empirical finding described as the "heritability paradox" is not really paradoxical at all. It is actually an example of a well-known effect in psychometrics — the *aggregation* of causal effects, whereby the sum or average of a number of correlated factors has greater reliability and generality than the average of the reliability coefficients of each of the aggregate's separate components. The factor common to a number of somewhat different but correlated ECTs, therefore, should be expected to have a higher phenotype–genotype correlation (and thus higher $h^2$) than its separate elements. Psychometric tests, even the relatively homogeneous subtests of a multi-test battery, are generally much more complex measures than are the ECTs typically used in measuring RT. A single psychometric test is typically composed of numerous nonidentical items each of which samples a different aspect of brain activity, even when the same elements in that activity are sampled by RT on a specific ECT that might consist of a virtually identical reaction stimulus on each and every test trial and therefore would elicit responses involving a much more restricted sample of neural processes. In view of the aggregation effect, it should also not be surprising to find that the composite RT on a number of varied ECTs shows values of $h^2$ that are very comparable to those found for most psychometric tests of intelligence and achievement. Achievements are also an aggregation, not just due to their variety of knowledge and skill content, but also because they are a cumulative, aggregated product of neural information processes occurring over extended periods of time. In acquiring knowledge and skills over an extended period of time, individual differences in speed of information processing are much like differences in compound interest rates acting on a given amount of capital. The seemingly slight but consistent individual differences in such relatively elemental processes as measured by ECTs, when acting over periods of months or years, therefore, can result in remarkably large individual differences in achievement. Individual achievements, reflecting largely the accumulated effects of individual differences in speed of information processing acting on cultural–educational input, therefore, generally show a greater genetic than environmental influence.

## Note

1.  A common error in the interpretation of the correlation between MZA is that the $h^2$ is the square of the MZA correlation, whereas, in fact, the MZA correlation itself is a direct estimate of $h^2$, as explained below:

    *Definitions*
    $P_1$, phenotype value of twin 1; and $P_2$, phenotype value of twin 2 (i.e., the co-twin of twin 1).
    $G_1$, genotype value of twin 1; $G_2$, genotype value of twin 2, and, as the twins are monozygotic, $G_1 = G_2 = G$.
    $E_1$, environmental value of twin 1; $E_2$, environmental value of twin 2.
    *The Model*

    $$P_1 = G + E_1 \text{ and } P_2 = G + E_2$$

    *Computation*
    The values $P$, $G$, and $E$ are in standardized score form, so that in a sample of twins the *means* of the values of $P$, $G$, and $E$ are all equal to 0, with SD (and variance) $= 1$.
    Assume the rearing environments of the separated MZ twins are uncorrelated, and that $G$ and $E$ are uncorrelated (i.e., random placement of co-twins in different environments).
    Then the correlation ($r_{12}$) between twins 1 and 2 in a large sample of $N$ pairs of twins (with members of each twin pair called 1 and 2) is

    $$r_{12} = \Sigma P_1 P_2 / N = \Sigma[(G+E_1)(G+E_1)]/N$$

    (*Note*: read $\Sigma$ as "the sum of ")
    Expanded, this becomes

    $$r_{12} = \Sigma G^2/N + \Sigma GE_1/N + \Sigma GE_2/N + \Sigma E_1 E_2/N$$

Since $G$ and $E$ are uncorrelated and $E_1$ and $E_2$ are uncorrelated, each of the last three terms in the above equation is equal to zero, and so these terms can be dropped out, leaving only

$r_{12} = \Sigma G^2/N$, which is the *genotypic variance* (or genetic variance, or heritability).

The present mean value of $r_{12}$ for IQ in all studies of MZ twins reared apart is .75.

The estimated population correlation between phenotypes and their corresponding genotypes, then, is $\sqrt{r_{12}}$. (The present best estimate of the phenotype-genotype correlation for IQ $= \sqrt{.75} = .87$.)

# Chapter 8

# The Factor Structure of Reaction Time in Elementary Cognitive Tasks

John B. Carroll, the doyen of factor analysis in the field of psychometrics, reviewed the existing literature in 1991 on the factor structure of reaction time (RT) and its relationship to psychometric measures of cognitive abilities (Carroll, 1991a). He factor-analyzed 39 data sets that might possibly yield information on the factorial structure of various kinds of RT measurements. His conclusion:

> My hope was to arrive at a clear set of differentiable factors of reaction time, but this hope could not be realized because of large variation in the kinds and numbers of variables that were employed in the various studies. The only rough differentiation that I could arrive at may be characterized as one that distinguished between very simple RT tasks, such as simple and choice RT, and tasks that involved at least a fair degree of mental processing including encoding, categorization, or comparison of stimuli. These two factors are sometimes intercorrelated, but sometimes not; the correlation varies with the type of sample and the types of variables involved. The main conclusion I would draw here is that the available evidence is not sufficient to permit drawing any solid conclusions about the structure of reaction time variables. (p. 6)

The one fairly certain conclusion that Carroll could glean at that time was that the correlation between RTs, on the one hand, and scores on conventional tests of cognitive abilities, on the other, originates in their higher-order factors, namely psychometric *g*. As he stated it, "...the correlation [between chronometric ECTs and psychometric tests] can be traced to the fact that the higher-order ability, in this case [psychometric] *g*, includes some tendency on the part of higher-ability persons to respond more quickly in ECT tasks" (p. 10). Later on, in his great work on the factor structure of cognitive abilities (Carroll, 1993), he was able to draw on newer studies that yielded further conclusions regarding the relationship between chronometric and psychometric measures, the subject of Chapter 9. The present chapter examines only the factor structure of RT in various ECTs separately from their correlations with psychometric tests. Unfortunately, this task is about as intractable today as it was for Carroll back in 1991, despite the greatly accelerated growth of RT research in recent years. The problem remains much the same as Carroll described it. Interest has been focused so much on the connection between RT and psychometric intelligence that the factor structure of RT tasks has never been systematically studied in its own right. So we are limited to the few possible generalizations about factor structure that can be gathered from the correlations among various ECTs originally obtained to answer different questions.

We are reasonably tempted therefore to ask: why bother with the factor analysis of RTs? Is asking about the factor structure of ECTs even the right question? The answers to these essentially methodological questions and the empirical conclusions that follow depend on being clear about certain factor-analytic concepts and their specific terminology.

## Factor Analysis and Factor Structure

Factor analysis is used to describe the structure of a given correlation matrix of $n$ variables in terms of a number of source traits, or *latent variables*, that cannot be directly measured but are hypothesized to explain the pattern of correlations among the $n$ observed variables. The factor analysis begins with a matrix of correlation coefficients among a set of directly measured variables, $V1$, $V2$, ... ,$Vn$, such as test scores. The computational procedure extracts from the correlation matrix a number of *factors* and *factor loadings,* representing the *latent traits* (hypothetical sources of variance) that mathematically account for the *structure* of the correlation matrix.

Factor analysis can be explained most simply in terms of a Venn diagram, shown in Figure 8.1. The total standardized variance of each of the three variables A, B, and C (e.g., scores on tests of verbal, perceptual-spatial, and numerical abilities) is represented as a circle. The standardized total variance, $\sigma^2 = 1$, of each variable is represented by the *area* encompassed within each circle. The shaded areas overlapping between any one variable and all the others represent the proportion of the total variance that the variable has in common with all of the other variables (termed the variable's *communality*, symbolized as $h^2$). The total of all the shaded areas (the sum of the communalities) is the *common factor variance* in the given set of variables. The *coefficient of correlation*, $r$, between any two variables is the square root of the total area of overlap between those two variables. The *non*overlapping area for any variable in the given matrix constitutes variance that is *unique* to the measurements of that particular variable. It is referred to as that variable's *uniqueness*, $U$, and is equal to $1-h^2$. Each and every measured variable has some degree of U,

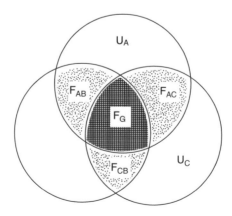

Figure 8.1: Venn diagram used to illustrate the gist of factor analysis.

which is composed of two parts: (1) *Specificity*, *S*, a true (i.e., reliable) source of variance that is not common to any other variable in the given matrix; and (2) random error of measurement, or unreliability (e).

The areas of overlap (shaded areas) represent *factors, F*, or common variance between two or more variables. In Figure 8.1 we see a total of three *F*s each of which has only two variables in common ($F_{AB}$, $F_{AC}$, $F_{CB}$). Because these factors comprise only particular *groups* of variables but not all of the variables, they are termed *group factors* (also called *primary factors* or *first-order factors*). One factor in Figure 8.1, $F_G$, is common to all of the variables and is referred to as the *general factor* of the given matrix. (The general factor should be labelled *G* for any matrix that does not contain a number of complex cognitive variables, such as IQ, that are typically considered the best exemplars of Spearman's *g*. The degree to which any obtained *G* factor resembles Spearman's *g* is a complex question that can only be answered empirically.)

In Figure 8.1 we see that the total variance comprising all three variables and their intercorrelations can be dichotomously divided in two ways: (1) *uniqueness* (*U*) versus *common factors* (all shaded areas), and (2) *group factors* versus a *general factor* ($F_G$). A variable's correlation with a particular factor is termed its *factor loading* on that factor. In Figure 8.1, the factor loading is the square root of the bounded area. Hence the square of a variable's loading on a given factor is the proportion of variance in that variable that is accounted for by the given factor. Factors are named according to the characteristics of the particular variables on which they have their larger loadings (termed *salient* loadings). It is especially important to note that a factor is definitely not an *amalgam* or average of two or more variables, but rather is a *distillate* of the particular source(s) of variance they have in common.

Although this Venn diagram serves to illustrate the gist of factor analysis, with only three variables, it is actually far too simple to be realistic. (For example, as a rule in factor analysis at least three variables are required to identify one factor.) But this simple Venn diagram can also be used to illustrate one other feature that is too often unrecognized in the use of factor analysis. When there is a significant *G* factor in the matrix, it should be clearly represented in the factor analysis of that matrix. The procedure known as *varimax rotation* of the factor axes is intended to maximize and roughly equalize the variance attributable to each of the group factors. This is fine if there is no *G* in the matrix. But if indeed the matrix actually harbors a *G* factor, varimax totally obscures it. The *G* variance is dispersed among the group factors in a way that makes them all perfectly uncorrelated with each other as well as inflating them with the variance that should rightfully be accredited to the *G* factor. This could be illustrated in Figure 8.1 by combining parts of $F_G$ with each of the group factors $F_{AB}$, $F_{AC}$, and $F_{BC}$ so as to maximize and equalize their variances as much as possible, at the same time maintaining zero correlations among the varimax factors. When a substantial *G* factor is revealed in the correlation matrix by any factor method properly capable of doing so, it is simply wrong to disperse its variance among the group factors. Varimax factors, however, may be useful in identifying the group factors in the matrix as a preliminary step to performing a hierarchical analysis.[1] But presenting a *varimax rotation* as a final result should be permissible only when one can first reject the hypothesis that the correlation matrix contains a *G* factor with significant loadings on every variable, a condition that is virtually assured for a matrix of all-positive correlations. In the domain of cognitive abilities this hypothesis has so far never been legitimately rejected (Jensen, 1998b, p. 117).

The term *factor structure* refers to a model that displays the structure of the correlation matrix in terms of its latent variables (factors). The three factor models most frequently encountered are the *hierarchical*, the *bi-factor* (or *nested*), and *principal factors* (with or without varimax rotation). These can be explained here without reference to the procedures for determining the number of factors to be extracted from a given matrix or the iterative computational procedures for obtaining the factor loadings. These and many other technical issues can be found in all modern textbooks on factor analysis. Factor analysis and components analysis are usually applied to a correlation matrix, but under special conditions, they can also be applied to a variance–covariance matrix.[1]

The common factors are sometimes referred to as *latent traits*, a term that implies nothing causal, but merely indicates that they are not directly observed or measured variables (e.g., test scores), but emerge from a factor analysis of the correlations among the directly observed variables. There are always fewer common factors than the number of variables, and the common factors comprise only some fraction of the total variance contained in the directly observed variables. Hence the measurement of factors in individuals is problematic; the obtained *factor scores* are always just *estimates* of the true factor scores, which can never be known exactly, however useful individuals' estimated factor scores might be theoretically or practically. (The most admirably clear and succinct nonmathematical discussion of the fundamental differences between factor analysis and principal components (PC) analysis I have seen is provided by the British statistician D. J. Bartholomew (2004)). PC analysis is sometimes used in place of factor analysis, usually omitting all components with latent roots (eigenvalues) smaller than 1. The result of a PC analysis looks similar to a factor analysis. But a PC analysis is technically not a latent trait model, as it analyzes the total variance including the uniqueness; therefore the components are "contaminated" by some admixture of uniqueness (i.e., specificity and error variance) and are therefore always a bit larger and a little less clear-cut than the corresponding common factors. (A correlation matrix of *n* variables contains *n* PCs, but usually in psychometric research only those PCs with the largest latent roots (typically eigenvalues >1) are retained in the final analysis.) Strictly speaking, PCs are not common factors, though they contain the common factors and are therefore usually highly correlated with them (Jensen & Weng, 1994). The most distinctly different models of common factor analysis and PCs are illustrated based on a matrix of nine intercorrelated variables. Examination of these factor matrices reveals the typical characteristics of each model.[2]

## Problems in Factor-Analyzing RT Variables

All of the factor analyses I have found that were performed exclusively of RT tasks are afflicted with one or more of the several known obstacles to attaining an optimal result. Specifically noting these deficiencies, to which purely psychometric analyses are also liable, helps to indicate the *desiderata* for future studies of the latent structure of RT in various ECTs. This could help in reducing redundancy in the measurement of RT variables, because the potential varieties of observed RT measurements undoubtedly far exceeds the number of latent variables.

(1) *The three variables rule.* The most conspicuously common violation in factor analyses of RT measures is the failure to apply the rule that each factor must be identified by at least three (or preferably more) different variables. If one hypothesizes the emergence of a particular factor, three different variables tapping fairly large amounts of that factor need to be included in the correlation matrix. The observed measurements of these variables cannot be merely equivalent forms of the same test; their high intercorrelations would not qualify as evidence for an authentic factor. The latent trait of interest must be manifested in more than a single test to qualify as a group factor. Otherwise only that part of its variance that is common to all other variables in the analysis will show up as a general factor, *G*; all of the remaining non-*G* variance constitutes the test's uniqueness, U. If a test actually harbors what could be a substantial *group* factor, that part of its variance remains hidden in the U variance as test *specificity*, S.

Such has been the fate of many RT variables in the factor analyses I have perused. These RTs appear to consist only of G and U, with small and nondescript loadings on any group factors. In some analyses there are no substantial group factors at all; only G and U emerge. We can identify authentic group factors only by including three or more variables that might possibly contain the same group factor. If a group factor emerges, it then can be further substantiated, or cross-validated, by identifying the same factor when the set of tests that originally identified it are included in a new correlation matrix containing tests other than those used in the original matrix. Such procedures have been used many times over to identify the main group factors in the enormous catalog of psychometric tests of cognitive abilities, which vastly exceeds the number of independent factors that account for most of their reliable covariation (Carroll, 1993).

(2) *Experimentally independent variables.* Another general rule in factor analysis is that the variables entering into it must be *experimentally independent*, that is, no two variables should be derived from the very same measurements. If A and B are experimentally independent variables in a given factor analysis, then any other variables containing A and B (e.g., A+B, A/B, etc.) should not be entered in the same correlation matrix. Examples of nonindependent measures are an individual's mean RT over *n* trials, the standard deviation (RTSD) of the individual's RTs, and the slope and intercept of RT, all derived from the very same data. Nonindependent measurements share identical measurement errors, which distorts their true intercorrelations. Such inflated correlations have entered into many factor analyses of RT data and at times can create the false appearance of a factor. The ideal way to solve this problem, if there is any good reason to include certain derived measures (e.g., individual means and SDs of RTs) in the same factor analysis is to divide each subject's RT performance into odd and even trials (e.g., measuring mean RTs on the odd trials and RTSDs on the even trials) thereby making the derived variables experimentally independent. This operation can be done twice, to utilize all of the data, by reversing the odd–even variables. The two correlations thus obtained can be averaged (using Fisher's *Z* transformation). Further, the correlation between the derived variables (e.g., RT and RTSD) can be corrected for attenuation using the average of the two odd–even correlations of the same

variable (boosted by the Spearman–Brown formula) as the reliability coefficient. The odd–even method is obviously not confounded by practice effects or intersession effects on the RT measurements. Note that there is no problem in including both RT and movement time (MT) in the same factor analysis, as they are experimentally independent measures.

### Practice Effects and Intersession Effects

The effects of these conditions have not been thoroughly investigated in factor-analytic studies, but there is good reason to believe that they can be manifested as "noise" in the intercorrelations among RT variables. Subjects show varying degrees of improvement in performance as they become more practiced on a given RT task or on a series of different ECTs in which subjects repeatedly use the same response console. There are intratask and intertask practice effects. Their magnitudes depend largely on task complexity, the more demanding tasks showing the larger practice effects. Because these effects may alter the factor composition of a task over the course of testing, they should be minimized by adequate practice trials and judicious sequencing of different tasks. Suitable data for factor analysis can be obtained from RT when enough practice trials have been given to stabilize not necessarily the absolute level of every individual's RTs *per se*, but the *reliability of individual differences* in RT across trials as indicated by the intertrial correlations. This reliability typically becomes fairly stable after relatively few (10–30) practice trials, even as the absolute RT continues to decrease gradually across a great many more trials, approaching asymptote at a decelerated rate. In using any new RT task it is necessary to run a parametric study to determine the course of changes in reliability of RT over $n$ trials as a guide to estimating the optimum number of practice trials for the particular task in a given population.

Even when individual differences in RT have stabilized in a given practice session, there is a rather marked loss of this stability across different test sessions on the same well-practiced tasks, when the same tasks are administered on different days or even a few hours apart on the same day. In other words, the intersession reliability of individual differences is lower than the intrasession reliability. Intersession and intrasession reliability coefficients on the same task have been found to differ mostly within the range of .70 to .95. For relatively simple RT tasks, such as the Hick paradigm, for example, we have found that intersession reliability remains rather consistently lower than intrasession reliability across 10 days of practice sessions. More complex tasks than SRT show greater intersession correlations (reliabilities), which increase over sessions, resulting in a matrix of intersession correlations that resembles a simplex (i.e., increasing correlations between successive sessions). Hence mixing a set of RT tasks administered in a single session together with tasks administered in different sessions may introduce sources of variance into the correlation matrix that can create puzzling inconsistencies in the factor loadings of various tests depending on whether they contain intersession variance, a *transitory* kind of *specificity*.

It should be noted in this context that such *transitory specificity* is, strictly speaking, not random measurement error, as it does not arise *within* sessions but arises only *between* sessions. It represents a true-score day-to-day fluctuation in an individual's RT, reflecting subtle changes in the physiological state that are generally not detectable with conventional psychometric tests. The differences in mean RT for an individual's "good" days may differ significantly from the mean RT on "bad" days. Such fluctuations reflect the great *sensitivity* of RT measures to slight changes in the individual's physiological state. Interestingly, the simpler RT tasks tend to be more sensitive to such fluctuations. The consistent magnitude of daily fluctuations in RT might itself be an important individual difference variable to be studied in its own right. Hence this sensitivity of RT should not be viewed as a drawback but rather as one of the advantages of RT for certain purposes, despite the correlational "noise" that such transitory specificity adds to a factor analysis. The solution to the problem, if one aims to achieve stable test intercorrelations and clearly interpretable group factor loadings, is to aggregate each individual's mean RTs on a given task obtained on two or more sessions, thereby averaging-out the transitory specificity due to intraindividual variation across sessions. If this may seem a Herculean labor for RT researchers, whoever said scientific research should be easy? We can compare our problems with the efforts of physicists to prove the existence of, say, neutrinos or antimatter. Our aim here is simply to try to identify reliably the main latent variables in the realm of those ECTs for which mean RT is in the range below 2 s for the majority of adolescents and adults (or 3 s for young children and the elderly). The much longer response latencies elicited by more difficult tasks often evoke strategies that make them factorially more complex and therefore less likely to line up with group factors reflecting more elemental processes. They tend to merge into one or more of the known psychometric group factors.

### *Excessive Task Heterogeneity*

This is probably the biggest problem in present factor analyses of ECTs. When multiple ECTs are used on the same group of subjects, theoretically nonessential task demands can be too varied for the latent traits to be clearly discernable. They are obscured by *method variance*. The response console, stimulus display screen, preparatory intervals, speed/accuracy instructions, criteria for determining the number of practice trials, and the like, should all be held constant across the essential manipulated variables in the given ECT. Different ECTs should vary only in the conditions that essentially distinguish one type of ECT from another in terms of their specific cognitive demands, such as attention, stimulus discrimination, or retrieval of information from short-term memory (STM) or long-term memory (LTM). Ideally, method variance should be made as close to nil as possible. Essentially, in designing a factor analytic study of ECTs one must think in terms of revealing their hypothesized latent variables, which dictates minimizing nonessential method variance in the observed RTs on a variety of different paradigms.

Also the different ECTs should be similar enough in complexity or difficulty level to produce fairly similar and preferably very low response-error rates. It is unduly problematic to deal statistically or factor analytically with large differences in error rates, either between different ECTs or between individuals on the same ECT. One solution is to retain only correct responses in the analysis. To ensure exactly the same predetermined number of error-free trials for every subject in the study sample, the sequence of test trials can be automatically programmed to recycle the stimulus that produced an error so it is to be presented again later in the sequence. Of course, all errors are recorded for other possible analyses. There should also be a predetermined criterion for discarding subjects who take an inordinate number of trials to achieve *n* error-free trials. Explicit rules for discarding outliers in a given study are seldom used in psychological research compared to their more routine use in the natural sciences. Outlier rules are often useful and occasionally necessary in mental chronometry.

Another category of outliers is the small minority of individuals whose performance on a given ECT fails to conform to the information-processing model it is supposed to represent (Roberts & Pallier, 2001). For example, in the Hick paradigm a few rare individuals do not show any linear increase or systematic change in RT as the number of bits of information increases. And one unusual subject (one of the world's greatest mental calculating prodigies) who was tested in my laboratory, although performing exceptionally well on the Sternberg memory scan task, showed not the slightest tendency to conform to the typical scan effect for this paradigm, i.e., a linear increase RT as a function of the set size of the series of a numbers; this subject's RTs were the same for all set sizes from 1 to 7 (Jensen, 1990). In such cases it seems apparent that the subject is actually not performing the "same task" as the great majority of subjects. Such atypical subjects should be treated as outliers, although they may be of interest for study in their own right. But their presence in a typical data set usually attenuates the ECT's correlations with other ECTs and psychometric variables.

### RT and MT Admixture Effects

This is a problematic kind of task heterogeneity. A response console using a home button permits the separate measurement of both RT and MT. But each measure, RT and MT, can contaminate the other one to some degree.

The problem in factor-analyzing response times to various ECTs that differ in type of response (single or double) is that RT and DT are not the same, as was explained in Figure 3.3. (p.53) RT and MT differ significantly even under otherwise completely identical conditions. The one study of this phenomenon, based on the Hick paradigm, is discussed in Chapter 3. (p.53) As hypothesized there, the "backward" effects of MT on RT are probably an example of Fitts's Law (Fitts, 1954) which states there is a monotonic relationship between the latency of a response calling for a particular movement and the complexity or precision for the required motor response. This implies that the specific movement has to be mentally programmed before it can be made, and that takes time, which is some fraction of the RT. Hence the RT in the double response condition requires more mental programming time than is required for an RT when a MT response is not required. Virtually all of the response programming time is included in the RT, while little, if any, of it gets into the MT, which itself is apparently a purely psychomotor variable rather than a cognitive one. And RT and MT seem not to be functionally related. In the Hick paradigm, for example, there is zero intraindividual (or *within* subjects) correlation between RT and MT, while there is a low but significant interindividuals (or *between* subjects) correlation (usually <.30) between RT and MT (Jensen, 1987b). The absence of an RT × MT correlation *within* subjects and the presence of a significant correlation *between* subjects generally suggests there is no directly causal connection between RT and MT. That is, although DT and MT are correlated in the population, they do not have a directly functional relationship in an individual.

Still unknown is whether the increment in RT that consists of the programming time required by the MT response is factorially the same or different from a theoretically pure RT.

### Information Processes versus Information Content

If the aim of measuring RT in various experimental variations of an ECT is to measure individual differences in a certain hypothesized process, such as speed of scanning information in STM, there is a question of how much of the individual differences variance is associated with the particular information *process* and how much is associated with the ECT's particular *content* (e.g., verbal, numerical, or figural). Unless these two sources of variance are experimentally manipulated in a *processes* ×*contents* design, a factor analysis of such data on a suitable sample of individuals would be uninformative as to the probable number of orthogonal factors needed to explain the correlation matrix or to estimate the proportions of individual differences variance attributable to each factor. The extant literature has not systematically addressed this question, and the little incidental evidence is still too meager and inconsistent to allow a confident answer. One large-*N* study, however, is suggestive and provocative (Levine, Preddy, & Thorndike, 1987). Three groups of various standardized psychometric tests known to measure verbal, quantitative, and visuospatial abilities were all correlated with six RT tasks in which the response stimuli of the ECTs consisted of either verbal, quantitative, or spatial materials. The result: the type of content made little difference in the correlations; the average correlations was $-.27$ when the content was the same for the RT and psychometric variables and $-.22$ when the content differed. A factor analysis of all the intercorrelations showed that a single general factor, $G$, common to both the RT and psychometric variables accounted for most of the common factor variance. The average loadings of the RT and psychometric variables on this general factor, $G$, are quite comparable ($-.40$ and $.43$). The RT variables had negligible loadings on psychometric group factors residualized from $G$. But there was also another factor, orthogonal to the general factor, with substantial loadings on only the RT variables. This factor probably reflects the strictly psychomotor component of RT, which is unshared by the nonspeeded psychometric tests. It is noteworthy that when the RT factor is residualized from the $G$ factor, its *largest* loadings are on the cognitively *least* demanding ECTs, such as SRT and 2-choice RT. (Also, psychometric tests have no nonzero loadings on the residualized RT factor.) It may seem unfortunate to some researchers if it is established that RT generally "reads through" the different contents of cognitive tasks, reflecting only their common factor independent of specific content. It would mean that, except for $G$, a greater variety of distinct cognitive factors are measured by psychometric than by chronometric tests. At present, however, it still remains to be determined by further investigation whether verbal, quantitative, spatial, and other psychometric group factors (residualized from $G$) can be measured chronometrically or if mental speed can only reflect their common factor.

### Is the Factor Analysis of Reaction Time Asking the Wrong Question?

The answer is Yes and No; it depends entirely on the question. It should be recalled that the correlation coefficients on which factor analysis is based are in turn based on the standardized deviations of individuals from the mean of the particular group that was tested and which is presumably a sample of some defined population. The linear correlation between any pair of variables measured in this sample, therefore, represents, *on average*, the relative deviations of each of the two variables from their respective group means. Knowing the coefficient of correlation between two given variables tells us the expected mean value of one test's *average* deviation (in standard deviation units) from its group *mean* given the other variable's average deviation from it's group mean. All the time we are dealing with group averages. A factor analysis of a correlation matrix, therefore, also can only reflect averages. Thus factors represent statistical entities, not individual entities, which can be estimated within this context only with a determinable margin of error.

The size of the correlations among various measures of cognitive variables indicates the *average redundancy* of these variables in the description of individual differences (variance) in abilities. In a factor analysis the redundancy among the variables is highlighted by the number of significant factors and also by the magnitudes of the variables' loadings in these factors. Redundant variables have highly similar loadings on each of the significant factors. They are, on average, interchangeable variables. In the extreme, it would be like measuring height both in centimeters and in inches. The factor analysis of chronometric measures is useful for selecting those tests that best measure different orthogonal (i.e., nonredundant) factors (in addition to *G*) and that have the largest factor loadings. These, then, are the tests that have the highest probability of yielding the most information in the particular population of interest. This is practically useful information and justifies the factor analysis of chronometric tests, in addition to their use in advancing basic research and theory on human cognition.

On the other hand, factor analysis can contribute little if anything to detecting or measuring exceptional conditions that are not detectably reflected by group means or variances in the general population. Yet many highly atypical and abnormal phenomena peculiar to a very few individuals are quite worthy of study. Though they would statistically be regarded as outliers in any population sample, they are entirely real and reliably measurable phenomena. An example in psychometrics is visual and auditory digit span memory, which are perfectly correlated in the normal population (hence they are factorially redundant measures). But they are poorly correlated among the rare individuals who have suffered a brain injury specifically affecting the auditory cortex. The infrequency of such individuals in the general population would preclude discovering the factorial separation of auditory and visual memory span by means of a factor analysis based on a sample of the general population. The same kind of differentiation of abilities under abnormal conditions could also occur in various chronometric measurements that might otherwise appear unitary or redundant in the factor analysis of typical population samples. The interpretation of RT measures in exceptional cases is considered later in reviewing the uses of chronometry for clinical diagnosis of abnormal brain conditions, monitoring their progression, and longitudinally assessing the effects of specific treatments and drugs in individual patients. This is probably the field of the potentially most valuable practical uses of chronometry.

### Empirical Evidence

Because of all the problems with the factor analysis of chronometric variables in the present literature, including the inordinate heterogeneity of the RT paradigms, apparatuses, and sample data, it is wholly unfeasible to attempt a true meta-analysis of all the findings. Nevertheless, there are still a few consistent observations that can be discerned from the extant data. As these generalizations apply to speeding up of processing measures regardless of whether they are based on DT or RT (as previously defined), from here on I will use RT as the generic term for all measures of response time based on a manual response.

### RT and MT are Distinct Factors

Although there is generally a slight but significant *positive* correlation between RT and MT, in factor analyses that include measures of both variables they load on two clear-cut orthogonal group factors. This is probably the most consistent and least ambiguous generalization that can be drawn from the whole factor analytic literature involving measures of RT and MT. Carroll's (1993) review of this evidence characterizes RT as a *cognitive* factor and MT as a *psychomotor* factor. Although worthy of study in its own right, MT is not considered in the subsequent discussion.

### Elementary Information Processes are not yet Identified as Group Factors

I have not found a factor analysis in which three or more distinct variables intended to measure one of the hypothesized elementary cognitive processes were included in the same correlation matrix with enough other RT variables to allow the possible emergence of factors representing the hypothesized elementary processes. Examples of the most well-known elementary processes are derived variables such as the slope of the Hick function (rate of information processing), the slope in the S. Sternberg paradigm (the rate of retrieval of information from STM), and the NI–PI measure (*name identity* RT–*physical identity* RT) from the Posner paradigm reflecting the rate of retrieval of information from long-term memory (LTM). The overall mean RTs obtained from these three paradigms do, however, have a large common factor. Whether derived measures of different elementary processes form distinct group factors when residualized from their common factor has not yet been determined.

### Significant Group Factors are Unidentified

Group factors orthogonal to *G* emerge in a few of the factor analyses and are large enough (in terms of the percentage of variance accounted for) to be considered significant factors. But there are two problems in trying to characterize them in psychological terms: (1) the few variables in which they are mainly loaded are too heterogeneous to reveal any common features that would provide a clue as to what precisely is the nature of the factor; and (2) there are too few large loadings to be able to identify a factor; the factor loadings are too similar in size to highlight any features that would afford a clue to what the factor common among the variables might be.

### A Large Common Factor, G, Exists in all Reaction Time Variables

I have found four studies containing independent data sets comprising between six and nine different RT variables in which procedures for the ECT measures within each study were methodologically fairly homogeneous. (The intercorrelation matrices for each of these four data sets are found in the following studies: Hale & Jansen, 1994; Kyllonen, 1985, Table 3, also reported in Carroll, 1993, Table 11.7; Miller & Vernon, 1996; Roberts & Stankov, 1999.) The specific aim here is to determine how many substantial factors can be extracted from each matrix and what percentage of the total variance in each matrix can be accounted for by these factors, with the remaining variance representing the variables' uniqueness. PCs analysis is probably best suited for this purpose. The first principal component (PC1) is interpreted as *G* and the other substantial components are the raw material for group factors residualized from *G* in a hierarchical factor analysis. (Remember that the chronometric *G* being discussed here is not necessarily Spearman's psychometric *g*. The relationship between *G* and *g* is the subject of Chapter 9.)

The criterion used here by which a PC is deemed "substantial" is based on the eigenvalues (latent roots) of the correlation matrix of *n* RT variables, using the Kaiser–Guttman rule for considering only those PCs with eigenvalues of 1 or greater than 1 as "substantial." The logic of this rule, simply, is that if one extracts PCs from the correlations among a large number of variables each consisting of purely random numbers (hence containing no true factors), all of the *n* extracted PC's eigenvalues hover around 1, the first one or two PCs are always slightly greater than 1 due to purely chance correlations in the matrix, and all the remaining components have very gradually decreasing eigenvalues of less than 1. (The sum of all of the eigenvalues is always equal to the number of variables, *n*. A particular PC's eigenvalue divided by *n* is the proportion of the total variance in all *n* variables accounted for by that PC.) When a typical correlation matrix is residualized from all of its PCs having eigenvalues > 1, the residualized correlation matrix typically looks just like a matrix of correlations among variables composed entirely of random numbers. Therefore, obtained PCs with eigenvalues smaller than 1 cannot be claimed to represent substantial latent variables.

The PC analyses of these four studies reveal two main points:

(1) All studies show a very large PC1 (or *G* factor). The PC1, on average, accounted for 57.4 percent (*SD* = 9.9 percent) of the total variance. This exceeds the percentage of variance accounted for by the PC1 of some standard psychometric test batteries, such as the Wechsler Intelligence Scales (about 40 percent). The chronometric *G* can only be interpreted at this point as *general speed of information processing*.

(2) In only one study (based on nine ECTs) was there a second component (PC2) that was substantial (with eigenvalue = 1.6), accounting for 17.8 percent of the total variance. (The PC1 accounted for 43.7 percent.) The PC2 is rather ambiguous but seems to contrast RTs for the simpler and more complex ECTs. In this study, in which the tasks were more liable to method variance than were the three other studies, the PC probably reflects different ratios of sensory-motor/cognitive abilities in the RT performance on the different types of ECTs.

Disregarding the ubiquitous sensory-motor aspect of RT, we are faced by a gaping question: can RTs on various ECTs reliably identify individual differences in *any other* cognitive latent traits besides a single common factor — the general speed of information processing?

The data provided by Hale and Jansen (1994) in the previous PC analyses seem to favor the hypothesis of an exclusive general factor accounting for all the reliable individual differences variance in different ECTs. This finding was explicated in a different context in Chapter 6 in relation to the high predictability of fast and slow groups' mean RTs on a considerable variety of ECTs, as shown in Figure 6.12.

The very same kind of plots was made for individual subjects selected from the high and low extremes and the middle of the total distribution of overall mean RT on all tasks. The resulting plots, shown in Figure 8.2, confirm the very same effect as seen in the group

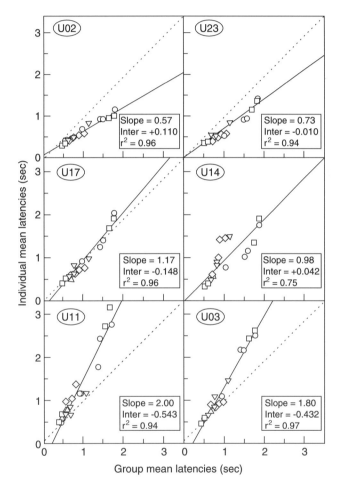

Figure 8.2: Plots for six individuals' RTs based on the same tasks shown as group means in Figure 6.12. Subjects were selected from the high and low extremes and the middle of the total distribution of RTs in a sample of 40 undergraduate university students. Note the high values of $r^2$ despite the marked differences in the slopes and intercepts of the regression lines. The one exception is subject U14. Retesting of this subject would determine if the exceptional deviations of the same data points represents a reliable individual differences phenomenon (from Hale & Jansen, 1994, reprinted with permission of the American Psychological Society).

mean RTs in Figure 6.12. That is, individual differences in all the various RT tasks differ along a single time dimension — general speed of information processing — while the differences in RT between the tasks reflect differences in task difficulty or complexity as represented by the slope measures in Figures 6.12 and 8.2. This finding, assuming it is strongly established in other RT data sets, thus resolves into two key questions: (1) What accounts for the *G* factor in RTs? and (2) What accounts for the reliable mean differences in RT across various ECTs? These remain open questions, partly because most of the RT data reviewed here were based on healthy, bright, young university undergraduates. A significantly different picture might result if the same kind of analyses were done in more heterogeneous population samples, especially if they included individuals with various kinds of brain pathology. In any case, if chronometric *G* is found to have substantial *external validity*, various measures of it would be important variables in their own right even if no other authentic speed of processing factors independent of *G* could be found.

# Notes

1. It should be noted that RT variables are particularly well suited to the factor analysis or principal components analysis of their raw-score variance—covariance matrix rather than the correlation matrix. The Pearson correlation coefficient is simply the standardized covariance, i.e., Cov $XY = [\Sigma (X - X) * (Y - Y)]/N$; Correlation $r_{xy} =$ Cov $XY/\sigma_x * \sigma_y$. It makes no sense to factor analyze a covariance matrix composed of raw-score variables that are not all on a scale with the same equal units of measurement. RT, being a ratio scale, which is the highest level of measurement, is one of the few variables in psychological research that could justify the use of covariances instead of correlations in a factor analysis or components analysis. But I have not come across an instance of its use in RT research. So I have tried factor analyzing RT data on a few covariance matrices to find out how the results differ from those obtained by analyzing the standardized covariances, i.e., correlations. Several points can be noted: (1) The factors that emerge are usually quite similar, but not always, as judged by the correlations of the column vectors of factor loadings and by the congruence coefficients between the factors. (2) The variables' factor loadings obtained from the covariance matrix reflect an amalgam of both (a) the factor structure obtained from the correlation matrix and (b) the differences in the variances of the variables. (3) In the covariance analysis, variables that have similar variance tend to load more similarly on the same factors. (4) In the covariance analysis, the first, most general factor strongly reflects different variables' similarity in variance. The loadings of variables on the general factor are, in a sense, weighted by the magnitudes of their variances. These features of covariance analysis may be most informative in the case of RT variables when looking for those particular variables that are related to an external criterion, such as IQ or other psychometric scores, because it is known that the RT tasks with greater individual differences variance are generally more highly correlated with other cognitive measures, particularly psychometric *g*. If one wants to obtain factor scores that would best predict performance on psychometric tests, therefore, the optimal method should be to obtain the factor scores actually as component scores from a PCs analysis of the RT

variables' raw covariance matrix. When a ratio scale with a true zero point is available, as in RT, some factor analysts (e.g., Burt, 1940, pp. 280–288) even go a step further and suggest factor analyzing the mean raw cross products of the variables, i.e. $(\Sigma XY)/N$. This brings the mean difficulty level of the tasks as well as their variances and intercorrelations simultaneously to bear in the results of the analysis, whether by factor analysis or component analysis.

2. Figure N8.1 illustrates the Spearman model in which one common factor is extracted from a set of variables (V1 – V9), with each variable loaded on a single factor (*g*) common to all the variables. Variance unaccounted for by the general factor is attributed to the variables' uniqueness (*u*).

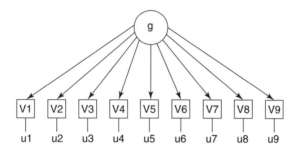

Figure N8.2 is the Thurstone model in which a number of uncorrelated factors (F1, F2, F3) are extracted. F1 may be a general factor, but if the factors are varimax rotated they remain uncorrelated (i.e., orthogonal factor axes) but the general factor variance is dispersed among all the common factors.

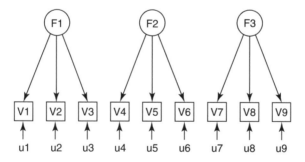

Figure N8.3 illustrates the bi-factor model (also called a *nested* model) in which a general factor is first extracted from the correlation matrix (as the first principal factor in a common factor analysis) and then the significant group factors are extracted from the variance remaining in the matrix. The group factors are uncorrelated because the general factor accounting for their intercorrelation was previously extracted. There is no hierarchical dependence of *g* on the group factors; because of this the *g* factor is always a fraction larger than the *g* extracted in a hierarchical analysis.

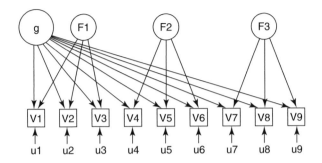

Figure N8.4 shows a hierarchical model in which the general factor is arrived at by first extracting group factors, which, if correlated with one another, allows a factor analysis of the group factors and the extraction of their common factor, *g*. In a matrix with very many variables there can be two levels of group factors, and so the general factor then emerges from the third level of the factor hierarchy. Factor loadings at each successive lower level of the hierarchy are residualized from the more general factors at the higher levels, creating an *orthogonalized* hierarchical structure in which every factor is perfectly uncorrelated with every other factor, thereby representing the correlations among all the measured variables in terms of a limited number of uncorrelated group factors.

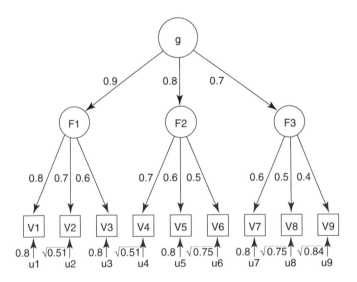

These models can be treated either as *exploratory factor analysis* (EFA) or as *confirmatory factor analysis* (CFA). CFA uses statistical tests of the goodness-of-fit of different factor models (EFA) to the data. Two or more different models are statistically contrasted against one another for their goodness-of-fit to the data in terms of their degrees of parsimony and conformity to certain theoretically derived expectations in explaining the correlational structure. The various models (except varimax when there is a large general factor)

yield highly similar results, typically showing very high (<.95) coefficients of congruence between the different models, particularly for the general factor in the domain of cognitive abilities (Jensen & Weng, 1994).

Table N8.1 shows examples of different models of factor analysis when each is applied to an analysis of the same correlation matrix. Table N8.2 shows the results of (1) a PCs analysis of the same correlation matrix used in Table N8.1, and (2) the varimax rotated components.

Table N8.1: Three factor models applied to the same correlation matrix.

| | **Hierarchical factor analysis** | | | | **Bi-factor analysis**[1] | | | | **Varimax rotation of principal factors** | | |
|---|---|---|---|---|---|---|---|---|---|---|---|
| | *g* | F1 | F2 | F3 | *g* | F1 | F2 | F3 | F1 | F2 | F3 |
| V1 | .72 | .37 | | | .74 | .29 | | | .70 | .30 | .25 |
| V2 | .63 | .31 | | | .66 | .23 | | | .61 | .26 | .22 |
| V3 | .54 | .26 | | | .57 | .18 | | | .53 | .22 | .19 |
| V4 | .56 | | .42 | | .59 | | .37 | | .24 | .63 | .18 |
| V5 | .48 | | .36 | | .52 | | .29 | | .21 | .54 | .15 |
| V6 | .40 | | .30 | | .44 | | .23 | | .17 | .45 | .13 |
| V7 | .42 | | | .43 | .44 | | | .41 | .17 | .15 | .54 |
| V8 | .35 | | | .36 | .37 | | | .33 | .14 | .12 | .47 |
| V9 | .28 | | | .29 | .30 | | | .25 | .11 | .10 | .38 |

[1]All factor loadings <.10 (constituting 0.50 percent of the total variance) are omitted.

Table N8.2: Principal components analysis and varimax rotation of the components based on the same correlation matrix used in the factor analyses in Table N8.1.

| | **Principal components** | | | **Varimax rotation of PCs**[1] | | |
|---|---|---|---|---|---|---|
| | PC1 | PC2 | PC3 | 1 | 2 | 3 |
| V1 | .77 | −.08 | −.31 | .76 | .28 | .20 |
| V2 | .72 | −.09 | −.37 | .76 | .21 | .15 |
| V3 | .64 | −.10 | −.47 | .78 | .11 | .09 |
| V4 | .66 | −.27 | .27 | .29 | .69 | .14 |
| V5 | .60 | −.29 | .34 | .21 | .71 | .11 |
| V6 | .52 | −.32 | .46 | .07 | .76 | .07 |
| V7 | .53 | .47 | .09 | .22 | .13 | .67 |
| V8 | .46 | .52 | .12 | .14 | .09 | .69 |
| V9 | .38 | .56 | .20 | .03 | .07 | .70 |

[1]Rotated PCs are technically no longer principal components (nor are they common factors as they contain uniqueness) and so are labeled as 1, 2, and 3.

# Chapter 9

# Correlated Chronometric and Psychometric Variables

By far the most extensive literature on the relationship between chronometric and psychometric variables is found in the study of mental abilities, particularly general intelligence. Although the earliest empirical studies in this vein date back at least as far as the research of Galton in the late nineteenth century, over 95 percent of the literature on reaction time (RT) and individual differences in mental ability has accumulated over just the past two decades.

The virtual hiatus in this line of research lasted for about 80 years. It is one of the more bizarre and embarrassing episodes in the history of psychology, and one that historians in the field have not adequately explained. A chronology of the bare facts has been outlined elsewhere (Jensen, 1982, pp. 95–98); Deary (2000a, pp. 66–72) provides the fullest account of the misleading secondhand reports of the early studies perpetuated for decades in psychology textbooks. It is a marvelous demonstration of how utterly deficient studies escape criticism when their conclusions favor the prevailing zeitgeist.

The classic example here is the often-cited study by Clark Wissler (1870–1947), a graduate student working under James McKeen Cattell, the first American psychologist to be personally influenced by Galton. The circumstances of this study, overseen by this eminent psychologist and conducted in the prestigious psychology department of Columbia University, could not have been more auspicious. Published in 1901, Wissler's study tested Galton's notion that RT (and various other sensory-motor tests) is correlated with intelligence. The result was a pitifully nonsignificant correlation of $-.02$ between "intelligence" and RT. The null result of Wissler's test on Galton's idea is what was most emphasized in three generations of psychology textbooks. What their authors seldom pointed out was that all the cards were outrageously (but naïvely) stacked in favor of the null hypothesis, for example: (1) the severe restriction of the range-of-talent in the subject sample (Columbia University students), which has the statistical effect of limiting the obtained correlation; (2) "intelligence" was not measured psychometrically but merely estimated from students' grade point average, which in selective colleges is correlated with IQ not more than .50; and (3) the reliability of the RT measurements (based on only three trials) could not have been higher than 0.20, as determined with present-day chronometers. Under such conditions, a nonsignificant result was virtually predestined. Yet for decades this study was credited with having dealt the heaviest blow against the Galtonian position! It remained the standard teaching about the relationship between RT and IQ until recently, apparently in total blindness to the fact that in 1904 a now historic classic by the English psychologist Charles Spearman (1863–1945) was published in the *American Journal of Psychology*, giving detailed notice of the methodological inadequacies of Wissler's study, and also introducing the statistical formula for correcting a correlation coefficient for attenuation (unreliability) due to measurement error.

When I began doing research on the correlation between RT and IQ, in the late 1970s, nearly every psychologist I spoke to about it was at best skeptical or tried to disparage and discourage the idea, in the firm conviction that earlier research had amply proved the effort to be utterly fruitless. Their annoyance with me for questioning this dogma was evident, despite my pointing out that I could find no valid basis for it in the literature. But I did find at least a dozen or so published but generally unrecognized studies (some reviewed by Beck, 1933) that made my venture seem a reasonably good bet. My friends' surprisingly strong conviction and even annoyance that my research program was taking a wrong turn decidedly increased my motivation to pursue the subject further. I was further encouraged by the revival of chronometry for the study of individual differences in the promising research of Earl Hunt and co-workers (1975) at the University of Washington and also that of Robert J. Sternberg (1977) at Yale University. At the time I sensed a changing attitude in the air that perhaps presaged a second chance for the role of mental chronometry in differential psychology.

But I have still often wondered why there was so strong an apparent prejudice against the possibility of finding that RT and mental speed are somehow related to intelligence. Why had this idea been resisted for so long by so many otherwise reasonable psychologists? The most likely explanation, I suspect, is the entrenchment of the following attitudes and implicit beliefs in many psychologists. These attitudes were bedrock in the psychological zeitgeist throughout most of the twentieth century.

(1) Any performance measurable as RT to an elementary task is necessarily much too simple to reflect the marvelously subtle, complex, and multifaceted qualities of the human intellect. A still pervasive legacy from philosophy to psychology is the now largely implicit mind–body dualism, which resists reductionist physical explanations of specifically human psychological phenomena. Any kind of RT was commonly viewed as a merely physical motor reaction rather than as an attribute of mind. Disbelievers in the possibility of an RT–IQ connection pointed out that many lower animals, for instance frogs, lizards, and cats, have much faster RTs than do humans (which in fact is true). And when confronted with good evidence of an RT–IQ correlation, they dismiss it as evidence for the triviality of whatever is measured by the IQ. These obstacles to research on RT are supported by belief systems, not by empirical inquiry.

(2) The speed of very complexly determined cognitive behavior is often confused with the sheer speed of information processing. It is noted, for example, that duffers at playing chess seldom take more than a minute or two for their moves, while the greatest chess champions, like Fisher and Kasparov, at times take up to half an hour or more to make a single move. Or it is pointed out that acknowledged geniuses, such as Darwin and Einstein, described themselves as "slow thinkers." Or that Rossini could compose a whole opera in less time than Beethoven would take to compose an overture. "Fast but superficial, slow but profound" is a common notion in folk psychology. But these anecdotes take no account of the amount or the "depth" of mental processing that occurs in a highly complex performance. The few times I have played against a chess master (and always lost), I noticed that all their responses to my moves were exceedingly quick — a second or two. But in tournament competition against others near their own level of skill, these chess masters typically take much more time in making their moves. Obviously they must have to process a lot more complex chess information when competing against their peers.

(3) Applied psychologists have resisted pursuing the RT–IQ relation mainly for practical reasons. There has existed no suitable battery of RT measures shown to have a degree of

practical validity for predicting external variables comparable to the validity of psychometric tests (PTs). Nor would RT tasks be as economical, as they require individual testing with a computerized apparatus with special software. So it is unlikely that RT tests could take over the many practical uses of standardized PTs, either individual or group administered. This is presently true. But RT methods have been conceived as serving mainly the purely scientific purpose of testing analytic hypotheses concerning the elemental sources of individual differences in the established factors of mental ability identified by complex PTs.

(4) Psychometricians have downgraded RT as a measure of cognitive ability because RT is mistakenly assumed to measure the same kind of test-taking speed factor that has been identified in PT batteries. This test-speed factor is observed in very simple tasks that virtually all subjects can easily perform. Individual differences in these highly speeded tests can be reliably measured only if the task is scored in terms of how many equally simple items the subject can execute within a short-time limit, such as 1 or 2 min, for example, the Digit Symbol and Coding subtests of the Wechsler Intelligence Scales. The common variance in these speeded tests typically shows up in a large factor analysis of various PTs as a small first-order factor with a weak relation to psychometric *g*. Its most distinguishing characteristic is its very low correlation with nonspeeded power tests, such as Vocabulary, General Information, and Similarities, or the Raven matrices. Various types of choice RT, on the other hand, have their highest correlations with the most highly *g* loaded nonspeeded PTs, and they show their lowest correlations with the speeded tests that define the psychometric speed factors, such as coding, clerical checking, and making *X*s, which have the lowest *g* loadings of any PTs. In this respect, the psychometric speed factor is just the opposite of RT measures. So the mistaken equating of mental speed as measured in chronometric paradigms with scores on highly speeded PTs has given the former a bum rap.

The idea that mental speed may be importantly involved in variation in human intelligence was not universally deprecated in American psychology. Early on, one of the pioneers of psychometrics, Edward L. Thorndike (1874–1949), the most famous student of J. McKeen Cattell, attributed a prominent role to mental speed in his set of principles for the measurement of intelligence (Thorndike, Bregmann, Cobb, & Woodyard, 1927). He also referred to these principles as the "products of intelligence." Within certain specified conditions and limits, all of these hypothetical generalizations have since been proved empirically valid. What is now needed is a unified theory that can explain each of them and the basis of their interrelationships. Because these five principles stated in Thorndike's *The Measurement of Intelligence* (1927) well summarize some of the most basic phenomena that need to be explained by a theory of individual differences in intelligence, they are worth quoting in full:

1. Other things being equal, the harder the tasks a person can master, the greater is his intelligence (p. 22).
2. Other things being equal, the greater the number of tasks of equal difficulty that a person masters, the greater is his intelligence (p. 24).
3. Other things being equal, the more quickly a person produces the correct response, the greater is his intelligence (p. 24).
4. Other things being equal, if intellect A can do at each level [of difficulty] the same number of tasks as intellect B, but in less time, intellect A is better. To avoid any

appearance of assuming that speed is commensurate with level or with extent, we may replace "better" by "quicker" (p. 33).

5. It is important to know the relation between level [difficulty] and speed for two reasons. If the relation is very close, the speed of performing tasks which all can perform would be an admirable practical measure of intellect. The record would be in time, an unimpeachable and most convenient unit (p. 400).

A year before the appearance of Thorndike's 1927 book, two psychologists at Harvard published a small study based on only five subjects. RT was reported to be correlated a phenomenal −.90 and −1.00 with scores on two intelligence tests. In retrospect, these correlations are recognized as obvious outliers — not surprising for such small-sample correlations. It is amazing that the study was not immediately repeated with a much larger sample! Nevertheless, the authors' conclusion was on target in noting the promise suggested by their experiment: "If the relation of intelligence (as the tests have tested it) to RT of any sort can finally be established, great consequences, both practical and scientific, would follow" (Peak & Boring, 1926, p. 94).

## Chronometric Correlations with Conventional Mental Tests

That fact that chronometric measures are correlated with scores on PTs of mental abilities is now firmly established. Presently, active researchers in this field have reviewed much of this evidence from various theoretical perspectives and have drawn remarkably similar conclusions (Caryl et al., 1999; Deary, 2000a, b; Detterman, 1987; Jensen, 1982, 1985, 1987a, 1998, Chapter 8; Lohman, 2000; Neubauer, 1997; Vernon, 1987). A true meta-analysis of all this evidence, however, is neither feasible, nor could it be very informative. The great heterogeneity of the subject samples, the obtained correlations, and the testing conditions of both the chronometric and psychometric variables calls for a different kind of summary of the empirical findings.

Overall, I estimate that less than 5 percent of the correlations reported in the RT–IQ literature are on the "wrong" side of zero, most probably due to errors of sampling and measurement. The vast majority of all the reported correlations between RT and mental test scores are *negative*, with their distribution centered well below the zero point on the scale of possible correlations. But of greater interest, theoretically, than the overall mean and SD of these correlations are the task conditions that systematically govern variation in their magnitudes.

There are many possible ways to categorize the relevant data. Probably the most informative from the standpoint of theory development are the following main types of quantitative relationship between RT tasks and PTs:

(a) Comparing the mean levels of selected high- and low-PT criterion groups on various measures of RT.
(b) Zero-order correlations between single RT tasks and single PTs.
(c) Multiple correlations between two or more RT tasks and a single PT, and vice versa.
(d) Correlations between latent traits, e.g., (i) canonical correlation between two or more RT and two or more PT variables; (ii) correlations between PT *g* factor scores and single RT

variables; (iii) factor analysis of a correlation matrix containing a variety of both RT and PT variables.

Nested in the above categories are other conditions that can affect the magnitude of the RT–PT correlation:

(a′) The particular kind of RT measurement (e.g., mean RT, reaction time standard deviation (RTSD), slope);
(b′) the specific content of the PT variable and the RT variable (e.g., verbal, numerical);
(c′) the particular elementary cognitive task (ECT) paradigm on which RT is measured;
(d′) the range of RT task difficulty, information load, or features of RT tasks that can be ordered on a "complexity" continuum; and
(e′) characteristics of the subject sample (range of ability, age, sex, educational level).

### Comparison of Criterion Groups

The simplest RT paradigm on which extensive ability group comparisons are available is the Hick paradigm. Figure 9.1 shows the mean RTs of three large groups of young adults of similar age selected from different regions of the IQ and scholastic achievement continuum. All were tested on the same RT apparatus under virtually identical conditions. An

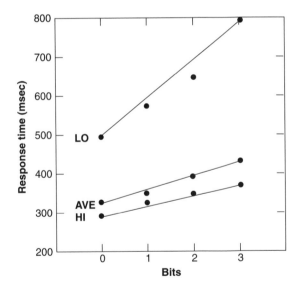

Figure 9.1: Mean RT in the Hick paradigm with 1-, 2-, 4-, and 8-choice RT tasks corresponding to information loads of 0, 1, 2, and 3 bits, in three young adult groups at three levels of ability labeled Hi (university students [N=588]), Ave (vocational students [N=324], and Lo (employees in a sheltered workshop [N=104]), with the respective groups' mean IQs of 120, 105, and 75. The intercepts (in milliseconds) and slopes (in milliseconds per bit) of the Hick function for each of the groups are: Hi, 295 and 25; Average, 323 and 35; Lo, 483 and 96. (Data from Jensen, 1987a.)

institutionalized group of 60 severely retarded adults (mean IQ about 40) was also tested on the Hick task with the same apparatus, but, on average, they did not conform to Hick's law, i.e., the linear regression of RT on bits (Jensen, Schafer, & Crinella, 1982). It is the only group reported in the RT literature that does not display Hick's law and also the only group for which movement time (MT) is *slower* than RT. However, in most groups of normal intelligence there are a few individuals who do not conform to Hick's law (see Jensen, 1987a, pp. 119–122). We also found that individuals with an IQ below 20 could not perform the Hick task, usually perseverating on the button used for the 0 bit condition regardless of the number of buttons exposed in subsequent trials. Nor was it possible to inculcate the Hick skill for the full range of information load (0–3 bits) in a half-hour training period.

These three criterion groups (in Figure 9.1) also differed on average in intraindividual variability (RTSD) in RT measured (in milliseconds) as the average standard deviation (SD) of RTs over 30 trials: Hi=37, Average =52, Lo =220. RTSD increases linearly, not as a function of bits (or the logarithm of the number ($n$) of possible alternatives of the response stimulus), but directly as a function of $n$ itself. RTSD is generally more highly related to intelligence differences than is RT (Jensen, 1992).

RT is also discriminating at the high end of the ability scale, as shown in Figure 9.2, comparing the mean RT of three criterion groups on each of the eight different ECTs that differ over a considerable range of difficulty as indicated by their mean RTs. The various RT tests are described elsewhere (Cohn, Carlson, & Jensen, 1985). The three criterion groups were 50 university undergraduates (Un); 60 academically gifted students (G), mean age 13.5 years, all with IQ above 130, enrolled in college level courses in math and science; and 70 nongifted (NG) junior high students in regular classes, mean age 13.2 years, scoring as a group 1 SD

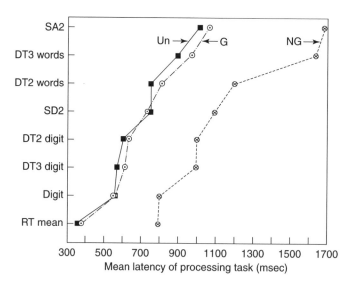

Figure 9.2: Mean latency (RT) on various processing tasks in three groups: university students (Un), gifted (G), and nongifted (NG) children (aged 12–14 years). (From Cohn, Carlson, & Jensen, 1985, with permission of Elsevier.)

above California statewide norms in scholastic achievement. The Un and G groups do not differ significantly in mean RT, but both differ about 1.3 SD from the NG group. Even on the simplest test (lifting the index finger from the home button when a single light goes "on"), the G and NG groups differ on an average by 54 ms. On all tasks, they differ by 700 ms, on average, a difference amounting to approximately 1.5 SD. Overall, the actual speed of processing for the NG group is 1.6 times slower than that of the G group. The profiles of mean latencies (RT) for each of the three groups, however, are highly similar, with correlations of .98 between each of the three pairs of profiles. Similar effects were found for measures of RTSD.

Another study compared academically gifted (G) and nongifted (NG) pupils, ages 11–14 years, differing 1.74 SD on Raven's Advanced Progressive Matrices test. Three easy RT tasks of increasing complexity were given: (1) simple reaction time (SRT), (2) 8-choice reaction time (CRT) (in the Hick paradigm), and (3) Odd-Man-Out (OMO) discrimination. They showed the following NG–G differences in mean RT (milliseconds) and RTSD (Kranzler, Whang, & Jensen, 1994).

|  | **SRT** | **CRT** | **OMO** |
|---|---|---|---|
| Difference in mean RT | 9 | 41 | 138 |
| Effect size ($\sigma$ units) | 0.15 | 0.78 | 1.24 |
| Difference in mean RTSD | 7 | 22 | 71 |
| Effect size for RTSD | 0.23 | 0.79 | 1.34 |

A Brinley plot is an especially revealing graphical method for contrasting different criterion groups simultaneously on a number of variables, because the "goodness-of-fit" of the data points (i.e., various tests) to a linear function reflects the degree to which the various tests are measuring a global factor that differentiates the criterion groups. The Brinley plot, of course, is meaningful only for ratio scale measurements. In one of the important articles in the RT literature, Rabbitt (1996) gives several examples of Brinley plots based on RT, each of them showing about the same picture for criterion groups that differ on PTs of ability. In one example, the subjects were 101 elderly adults (aged 61–83). Two criterion groups, closely matched on age, were selected on the Cattell Culture Fair (CF) Intelligence Test. The low IQ group had CF raw scores between 11 and 29 points; the high group had CF scores between 29 and 40. The low and high CF means differ about 3?. In the Brinley plot of these data, shown in Figure 9.3, the bivariate data points for 15 simple tasks with mean RT ranging between 200 and 1750 ms fall on a regression line that accounts for 99 percent of their variance. As the ECTs increase in information processing load, the differences between the high and low CF groups increase by a single global multiplicative factor. This factor differs between individuals, strongly affecting processing speed on every test. The larger the task's information load, the greater is the total processing time by a constant factor for each individual — a factor clearly related to IQ. But it is also instructive to consider Figure 9.4, which shows the relationship between the mean RTs on these 15 ECTs and the magnitudes of their Pearson correlations with the CF IQ. The fact that the data points do not fall very close to the regression line (accounting for $r^2 = .56$ of the variance) indicates that some other factor(s) besides sheer processing speed (for instance, working memory) are probably involved in the ECTs' correlations with IQ.

Essentially, the same effect is found for Brinley plots that compare overlapping groups of young adults of average (*N*=106) and high (*N*=100) ability (mean IQs ≈ 100 and 120) tested on seven ECTs that vary in information load (Vernon, Nador, & Kantor, 1985). Plots are shown for both mean RT and mean RTSD in Figure 9.5. Though RTSD shows a very large multiplicative effect, less of its variance is explained by the linear regression than in the case of mean RT.

## Correlations of Single Chronometric and Psychometric Variables

Typically, the smallest chronometric–psychometric (*C–P*) correlations are found between single *C* and *P* variables. This is particularly true for single *C* variables because they are usually much more homogeneous in content and cognitive demands than are *P* tests, which are invariably composed of a considerable number of varied items scored right or wrong for which the total score reflects a considerably broader factor than the score on a single *C* test. Increasing the number of trials given on a particular *C* test increases the reliability of measurement of the subject's mean (or median) RT and RTSD, but it has either no effect or a diminishing effect on the breadth of the factor it measures.

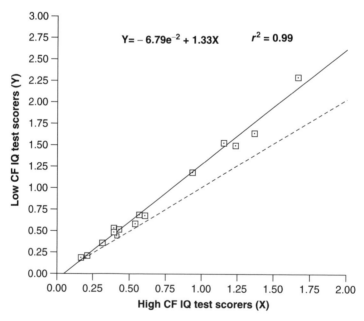

Figure 9.3: A Brinley plot of response time measures (in seconds) on 15 tasks given to adult groups in the lower (Low IQ) and upper (High IQ) halves of the distribution of scores on the CF Intelligence Test. The data points are well fitted by the linear regression ($r^2 = .99$). The dashed line (added by ARJ) is the hypothetical regression line assuming the Low IQ group performed exactly the same as the High IQ group. (From Rabbitt, 1996, with permission of Ablex.)

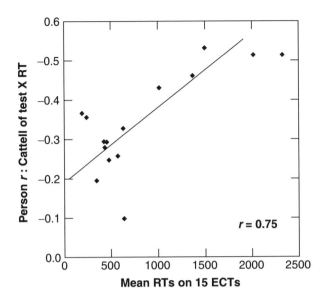

Figure 9.4: A correlation scatter diagram showing the relationship (*r*=.75) between the mean RTs on 15 ECTs and these ECTs' correlations with the Cattell CF IQ. (Graph based on data from Rabbitt, 1996, p. 82, Table 1.)

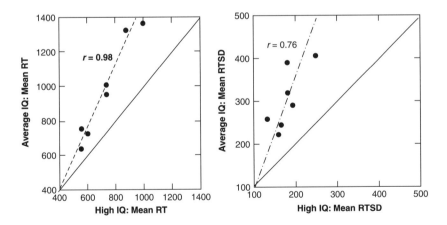

Figure 9.5: Brinley plots comparing High IQ and Average IQ groups on mean RT (left panel) and mean RTSD (right panel) for seven diverse ECTs. The solid line represents the hypothetical regression line assuming the Average IQ group performed the same as the High IQ group. (Graphs based on data from Vernon, Nador, & Kantor, 1985, pp. 142–144, Tables 1 and 3.)

The following brief collection of examples is not intended to be a comprehensive review. An assortment of examples was selected from a wide variety of ECTs so as to give a fair idea of the magnitudes and variability of the single correlations found between single *C* and *P* variables. Much of the variability in *C–P* correlations, even for the same ECT paradigm, is associated with differences in the range of psychometric ability within the different subject samples. How much variability in *C–P* correlations for nominally the same ECT is associated with differences in laboratory equipment, procedures, and test conditions is unknown; these variables have not yet been systematically studied.

### *Hick Paradigm*

The simplicity of the Hick paradigm makes it a good example. Beyond the 0 bit condition (SRT) it is a set of CRTs usually based on 2, 4, and 8 possible choices, which are scaled as 1, 2, and 3 bits. The simplicity of the tasks is shown by the fact that for the hardest condition (3 bits) normal adults have RTs of less than 500 ms. Figure 9.6 shows the unweighted and *N*-weighted mean correlations between RT and IQ as a function of bits within 15 independent groups that are each considerably more homogeneous in IQ than is the general population (Jensen, 1987a). The correlations are all quite low, but it is of theoretical interest that they are a linear function of bits. The *N*-weighted and unweighted correlations between IQ and the aggregated RT over all bits, corrected for both restriction of range and attenuation, are $-.31$ and $-.39$, respectively. This is about as close as we can come to estimating the population RT–IQ correlation in the Hick paradigm (Jensen, 1987a). (The corresponding RTSD–IQ correlations are $-.32$ and $-.42$.) When Hick data were obtained with a quite different apparatus in a sample of 860 U.S. Air Force enlistees, the RT–IQ correlations were slightly higher than those in Figure 9.6, ranging from $-.22$ to $-.29$, but surprisingly in this sample there was no significant trend in the size of the correlations as a function of bits (Detterman, 1987). The IQ–RT correlation for the combined RT data, corrected for

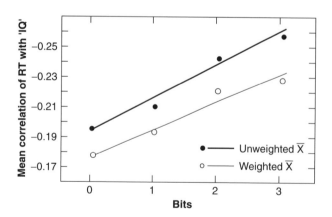

Figure 9.6: Unweighted and *N*-weighted mean correlations between RT and IQ as a functions of bits in the Hick paradigm, based on 15 independent samples with total $N = 1129$. (From Jensen, 1987a, p. 164, with permission of Ablex.)

restriction of range in the Air Force sample, was −.47, which can be regarded as another statistical estimate of the population correlation.

In a very heterogeneous group with Wechsler Performance IQs ranging from 57 to 130, the RT–IQ correlations were especially large and there was a strong linear trend in RT–IQ correlations as a function of bits, shown in Figure 9.7. Although the distribution of IQs in this specially selected sample does not represent any intact population group, being more a rectangular than a normal distribution. But it strongly confirms the phenomenon of the Hick RT–IQ correlation and its linear regression on task complexity scaled as bits.

In a truly random population sample of 900 middle-aged adults (age around 56 years) in Scotland, the observed RT–IQ correlations for simple RT and 4-choice RT were −.31 and −.49, respectively. This sample included the full range of IQs in the population, the distributions of both RT and IQ were close to normal (perfectly Gaussian for the middle 95 percent of the distribution), and the RT–IQ *correlations* did not vary significantly between subgroups defined by age, social class, education, or error rates in the 4-choice RT (Deary, Der, & Ford, 2001; also see Der & Deary, 2003 for further analyses of these data). Given this excellent population sample, it is especially instructive for a theory of the RT–IQ correlation to view the particular form of the relationship between SRT and CRT in relation to IQ (reported in terms of deciles of the total distribution on the Alice Heim 4 IQ test), shown in Figure 9.8. Note especially the linear regression of CRT on SRT. It indicates that CRT is a constant multiple of SRT as a function of IQ. (The goodness-of-fit to linearity would probably be even better for a measure of *g*.) The same near-linear trend of CRT as a decreasing function of IQ, as shown in Figure 9.8, was also found for each of the

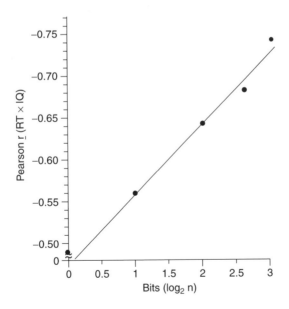

Figure 9.7: Correlations between choice RT and IQ as a function of the number of choice alternatives scaled in bits, in a group of 48 persons with Wechsler Performance IQs ranging from 57 to 130. (Data from Lally & Nettelbeck, 1977.)

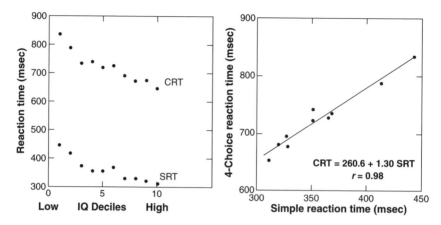

Figure 9.8: *Left*: SRT and CRT plotted as a function of IQ in deciles from lowest (1) to highest (10). *Right*: CRT plotted as a function of SRT. (Data from Der & Deary, 2003, Table 2.)

three different CRT tasks in a group of college students, all of them in the upper-half of the population-normed distribution on the CF IQ test (Mathews & Dorn, 1989, p. 311).

### Memory Scan (MS) and Visual Scan (VS) Tasks

These are also known, after their inventors, as the Sternberg (1966) and the Neisser (1967) paradigms. In the MS task, a series of 1–7 digits (called "the positive set") is presented for a brief time (3 s), then a single digit appears to which the subject responds as quickly as possible by pressing one of the two buttons labeled YES and NO, indicating whether the single number *was* or *was not* a member of the positive set. The mean RTs increase as a linear function of set size. In the VS task, the order of stimulus presentation is just the reverse of that in the MS task. First, a single digit is presented (hence making little demand on memory). Then a set of 1 to 7 digits is shown, which the subject visually scans and responds YES or NO according to whether the single digit is present or absent in the scanned set. The RTs for MS and VS hardly differ, and individuals' RTs on MS and VS are correlated .91 (disattenuated $r=1.00$) (Jensen, 1987b). This indicates that one and the same latent variable is involved in individual differences on each of these rather dissimilar paradigms, which differ mainly in their demands on short-term memory (STM).

The average correlations of individual differences in mean (or median) RT (and derived parameters) on MS and VS with "IQ" (i.e., scores on various intelligence tests) averaged over five independent studies totaling 314 subjects are shown in Table 9.1. The mean RT and RTSD consistently show rather low but significant correlations with psychometric intelligence, whereas the derived measures, intercept and slope, of the regression of RT on set size are virtually zero and some are even on the theoretically "wrong" side of zero. Theoretically, of course, the intercept represents only the sensory-motor component of RT (in effect, RT for set size = 0), rather than any cognitive processing component, so a zero correlation with IQ is not expected. In reality, however, SRT does have some cognitive component because of uncertainty (i.e., bit > 0) of the time of onset of the RS.

The nonsignificance of the slope parameter, however, is obviously theoretically troubling for any theory that posits processing speed as an important component of intelligence. However, as explained elsewhere (Jensen & Reed, 1990), because of lower reliability and statistical artifacts the RT slope is severely handicapped as a measure of *individual* differences. But it is not necessarily true for *group* differences. Because measurement errors average-out in the means of fairly large groups, it would be of critical theoretical value to determine if there is a significant difference between high IQ and low IQ groups in the overall slopes of the RT means of MS and VS regressed on set size. I have not found such a study.

### Semantic Verification Test

The SVT is intended as a simplified version of the various sentence verification tests that have been used in linguistic and cognitive research. It is fully described elsewhere (Jensen, Larson, & Paul, 1988). Instead of using full sentences to describe a simple stimulus, such as "The star is above the cross," the present SVT uses single prepositions, such as BEFORE, AFTER, FIRST, LAST, BETWEEN (and their negation, NOT BEFORE, etc.), to describe the position of one letter among a set of three. The three letters are always one of the six permutations of the letters **A B C**. First, the subject presses the home button on a binary response console; then, presented on a computer monitor for 3 s is a statement such as C after B. The screen then is blank for 1 s and the letters **A B C** appear on the screen, and the subject responds as quickly as possible by moving the index finger from the home button and pushing either the YES or the NO button (in this example the correct response is YES). RT is the time interval between the onset of the reaction stimulus (RS) **A B C** and the subject's releasing the home button. The mean difficulty levels of the various permutations of SVT items differ markedly and consistently. For university students, the average RTs on the different permutations range from about 600 to 1400 ms; and RTs to the negative statements average about 200 ms longer than to the positive statements (Paul, 1984).

Even in the restricted range of IQ for Berkeley undergraduates there is a correlation of $-.45$ between the total RT on the SVT and scores on the Raven Advanced Progressive Matrices (on which the subjects were told to attempt every item and to take all the time

Table 9.1: Correlations of various RT parameters with "IQ" in memory scan and visual scan tasks.[a]

| RT parameter | Memory | Visual |
|---|---|---|
| Mean/Med RT | $-.293$ | $-.266$ |
| RTSD | $-.279$ | $-.289$ |
| Intercept | $-.169$ | $+.060$ |
| Slope | $+.056$ | $+.016$ |

[a]Average correlations from five independent studies (Jensen, 1987b, Table 14).

they needed). There are also large mean differences on the SVT between university undergraduates and Navy recruits (Jensen et al., 1988). The SVT was taken by 36 gifted (IQ>135) seventh graders and their 36 nearest-in-age siblings. The two groups differed $1.09\sigma$ in IQ (Raven Advanced Progressive Matrices). Their mean RTs on the SVT differed almost as much as $-0.91\sigma$. The RT–IQ correlation in the combined groups was $-.53$; the $r$ was $-.43$ in the gifted group and $-.42$ in their siblings. The corresponding correlations for RTSD are $-.60$ and $-.47$ (Jensen, Cohn, & Cohn, 1989). It is worth noting that MT showed no significant difference between the gifted and sibling groups and no significant correlations with IQ in either group or the combined groups, again suggesting that MT is not a cognitive or $g$-loaded variable. The unique theoretical importance of this sibling study, however, is that it controls the environmental between-families background factors (social class, income, ethnicity, general nutrition, etc.) and yet shows that the RT–IQ correlations remain significant both within-families (i.e., among siblings) and between-families. In other words, the population correlation between RT and IQ does not depend on differences in the kinds of environmental variables that systematically differ between families.

### Coincidence Timing

Coincidence timing (CT) is one of the simpler types of chronometric measurement. However, it cannot be considered as SRT, because the subject's response reflects not only the speed of stimulus apprehension, as does SRT, but also calls for further information processing in the anticipation and prediction required by the CT task. In this procedure, the subject, with a finger poised on the response key, views a computer monitor on which there is a stationary vertical line extending from the top center to the bottom center of the screen. Then a small square enters the screen from either the right or left side, traveling in a straight line horizontally (or in a random path on one-third of the trials) at a constant speed of 10 cm/s. The subject is instructed to press the button exactly when the small square coincides with the vertical line. Two basic scores are derived from the subject's performance: the mean and SD over trials of the *absolute distance* (i.e., error) between the small square and the vertical line at the time that the subject pressed the response key. In a group of 56 eighth-grade students (mean age 13.5 years), the mean and SD of the CT error scores were significantly correlated with IQ (Standard Progressive Matrices) $-.29$ and $-.36$, respectively. Curiously, when the effect of the subjects' sex is statistically removed, these correlations are increased slightly to $-.36$ and $-.37$; the correlation with IQ of a combined score of the CT mean error and its SD over trials was $-.40$. All of these correlations have significance levels of $p<.003$ (Smith & McPhee, 1987).

### A Variety of Single and Dual Task Paradigms

RT tasks that make greater processing demands generally have longer RTs and larger correlations with IQ tests and particularly with $g$, the latent factor common to all such tests. For example, some tasks, like the Hick paradigm, require no retrieval of information, whereas paradigms like the Sternberg memory-scanning task call for the retrieval of information from STM, and the Posner Name Identity–Physical Identity task calls for retrieval of information from long-term memory (LTM). The retrieval process takes time. One way to experimentally manipulate the cognitive demands on RT tasks is by requiring the subject to perform two tasks within a brief time period. For example, incoming information has to be momentarily held in STM while performing a second and different processing task after which the information in STM has to be retrieved. This is called a *dual task* paradigm (described in detail in Chapter 2, pp. 31–32). Studies that included a battery of both single and dual task RT paradigms are described in Chapter 6 (pp. ). Besides the Hick task, the other seven tasks consist of the Sternberg and Posner paradigms, which are presented both as single tasks and, in certain combinations, as dual tasks.

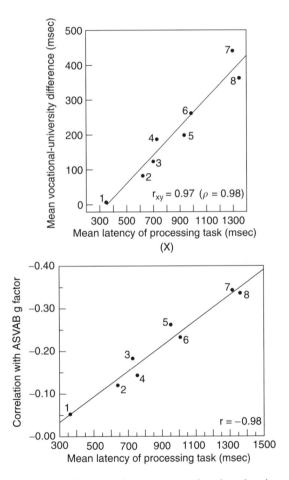

Figure 9.9: Left panel: RT differences between vocational and university students as a function of task difficulty as indexed by mean latency (RT) on each task in the combined groups. Right panel: Correlations of eight RT tasks with ASVAB *g* factor scores as a function of task difficulty. Single tasks are numbers 1, 2, 5, 8; dual tasks are numbers 3, 4, 6, 7. (From Vernon & Jensen, 1984, with permission of Ablex.)

The panel on the left side in Figure 9.9 shows the mean differences in RT between groups of vocational college (V) and university (U) students on the eight processing tasks (Vernon & Jensen, 1984). Task difficulty is indexed by the overall mean latency (RT) of the task in the combined groups. The V–U differences are closely related to the tasks' processing demands as indicated by their mean latencies. The panel on the right in Figure 9.9 shows the correlations of task difficulty (indexed by the mean latency) of each of the tasks with *g* factor scores obtained from the 10 subtests on the Armed Services Vocational Aptitude Battery (ASVAB). Although the single correlations between individual differences in RT and *g* factor scores (i.e., the ordinate in the right panel of Figure 9.9) are all smaller than − .40, the correlation between the V–U mean differences in RT and the tests' *g* loadings is .95, indicating that variation in the magnitudes of the V–U mean RT differences is almost entirely a result of the group difference in psychometric *g*. The *g* factor is manifested in RT performance to the extent of the task's cognitive demands.

# Multiple and Canonical Correlations

## *Multiple Correlation*

A multiple correlation coefficient (*R*) yields the maximum degree of liner relationship that can be obtained between two or more independent variables and a single dependent variable. (*R* is never signed as + or −. $R^2$ represents the proportion of the total variance in the dependent variable that can be accounted for by the independent variables.) The independent variables are each optimally weighted such that their composite will have the largest possible correlation with the dependent variable. Because the determination of these weights (beta coefficients) is, like any statistic, always affected (the *R* is always inflated) by sampling error, the multiple *R* is properly "shrunken" so as to correct for the bias owing to sampling error. Shrinkage of *R* is based on the number of independent variables and the sample size. When the number of independent variables is small (<10) and the sample size is large (>100), the shrinkage procedure has a negligible effect. Also, the correlations among the independent variables that go into the calculation of *R* can be corrected for attenuation (measurement error), which increases *R*. Furthermore, *R* can be corrected for restriction of the range of ability in the particular sample when its variance on the variables entering into *R* is significantly different from the population variance, assuming it is adequately estimated. Correction of correlations for restriction of range is frequently used in studies based on students in selective colleges, because they typically represent only the upper half of the IQ distribution in the general population.

Two examples of the multiple *R* between several RT variables and a single "IQ" score are given below. To insure a sharp distinction between RTs based on very simple ECTs and timed scores on conventional PTs, the following examples were selected to exclude any ECTs on which the mean RTs are greater than 1 s for normal adults or 2 s for young children. Obviously, not much cogitation can occur in this little time.

The simplest example is the Hick paradigm. Jensen (1987a) obtained values of *R* in large samples, where the independent variables are various parameters of RT and MT derived from Hick data, viz. mean RT, RTSD, the intercept and slope of the regression of RT on bits, and mean MT.

Without corrections for attenuation and restriction of range in the samples, $R=.35$; with both of these corrections, $R=.50$. This is the best estimate we have of the population value of the largest correlation that can be obtained between a combination of variables obtained from the Hick parameters and IQ as measured by one or another single test, most often the Raven matrices.

Vernon (1988) analyzed four independent studies totaling 702 subjects. Each study used a wide variety of six to eight ETCs that were generally more complex and far more heterogeneous in their processing demands than the much simpler and more homogeneous Hick task. The average value of the multiple $R$ (shrunken but not corrected for restriction of range) relating RT and IQ was .61, and for RTSD and IQ $R$ was .60. For RT and RTSD combined, $R$ was .66.

### *Canonical Correlation*

This might be regarded as the simplest form of a latent trait model. Instead of determining the correlation between observed variables (e.g., test scores), a canonical correlation (CR) calculates the correlation between (a) the common latent trait(s) in a given set of two or more observed variables and (b) the common latent traits(s) in another set of two or more observed variables. It is like a multiple $R$ in which *both* the independent variables and the dependent variables consist of a number of different measurements. However, the $C$ divides the common variance between the two sets of variables into orthogonal (i.e., uncorrelated) components, called canonical variates. All of the $C$s I have found between chronometric and psychometric variables have only one significant canonical variate; this first canonical variate, however, is quite substantial and it is the common factor linking both sets of variables. Like $R$, the $C$ is always an absolute (i.e., unsigned) value. Like $R$, $C$ is also inflated by random sampling errors and therefore is usually "shrunken" to correct for this statistical bias.

The $C$ model of the common latent variable in sets of psychometric and chronometric variables that represents the highest significant correlation that can be obtained between the two sets of variables is illustrated in Figure 9.10, from data on 96 college sophomores,

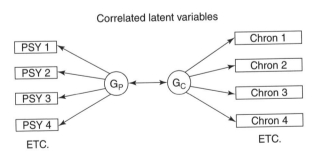

Correlated latent variables

Figure 9.10: A canonical correlation ($C$) between four chronometric tests and four PTs illustrates the method applied to correlational data from a study by Saccuzzo et al. (1986) described in the text.

in which the general chronometric variate ($G_C$) of four chronometric measures and that of four psychometric variables ($G_p$) have a $C=.65$ (shrunken $C=.55$). The correlation of each of the variables with the common factor of each set is shown in each of the boxes. The chronometric measures were simple RT, 3-choice RT, 5-choice RT, and inspection time (IT). The psychometric measures were the Wechsler Vocabulary and Block Design subtests, the Scholastic Aptitude Test, and the freshman grade point average (Saccuzzo, Larson, & Rimland, 1986).

Kranzler and Jensen (1991) obtained the $C$ in much larger sets of chronometric (37) and psychometric (11) variables. The chronometric variables were the RT, RTSD and MT, MTSD on the following paradigms: simple RT, 8-choice RT, and OMO RT, Sternberg memory scan, Neisser visual scan, Posner binary RT for same–different words and on synonyms–antonyms, and IT. The psychometric variables were the Raven Advanced Progressive Matrices and the 10 subtests of the Multivariate Aptitude Battery. In a sample of 100 university undergraduates, the $C$ between these two sets of variables was .60 (shrunken = .57). After correction for restriction of range in the university sample (with mean IQ of 120), the shrunken $C$ was .72, a value about the same as that of the average correlation between various IQ tests (Jensen, 1980b, pp. 314–315).

In a sample of 109 children (aged 4–6 years), Miller and Vernon (1996) obtained RT data on eight diverse ECTs on each of which the RT averaged less than 2 s; the error rate was 8 percent. The psychometric variables were the 10 subtests of the Wechsler Preschool and Primary Scale of Intelligence (WPPSI). The uncorrected $C$ between the RT and WPPSI batteries was .65.

Overall, the several estimates of the canonical correlations between sets of diverse psychometric mental ability tests and sets of various measures of RT in the various studies range from .55 to .72, averaging .62.

## Factor Analysis of Chronometric and Psychometric Variables Together

This is the most analytic method of looking at the relationship between the two classes of measurements. In terms of the number of RT measures (both direct and derived) obtained from several ECTs and the number of PTs, the factor analysis by Carroll (1991b) of the data from Kranzler and Jensen (1991), described in the previous section (p. 171), is probably the most revealing. Carroll performed Schmid–Leiman orthogonalized hierarchical factor analysis of these data, which comprised correlations among 27 chronometric and 11 psychometric variables. The hierarchical structure of the factor matrix is quite complex, but the main gist of it can be most easily explained in terms of the simplified schematic diagram shown in Table 9.2. Before doing the factor analysis, Carroll properly reflected all of the correlations between the chronometric and psychometric variables so that "goodness" of performance on both types of variables is represented by a positive (+) correlation between them. This resulted in positive signs for all of the salient or significant factor loadings in the whole analysis. All of the PT and all of the ECT:RT variables (but not the ECT:MT variables) have salient loadings on a large common factor labeled as $g$ here . But also note that the PT and ECT variables have no common group factors. Besides their loadings on $g$, the PTs load only on the PT group factors Verbal and Spatial. The ECTs load on

Table 9.2: Orthogonal Factors.

| Variable | g | Psychometric | | | Chronometric | | | |
|---|---|---|---|---|---|---|---|---|
| | | **V** | **S** | **M** | **RT** | **RTS** | **RTNS** | **MT** |
| Psychometric Tests | + | + | | | | | | |
| | + | + | | | | | | |
| | + | | + | | | | | |
| | + | | + | | | | | |
| | + | | | + | | | | |
| | + | | | + | | | | |
| Chronometric Tasks—RT | + | | | | + | + | | |
| | + | | | | + | + | | |
| | + | | | | + | + | | |
| | + | | | | + | | + | |
| | + | | | | + | | + | |
| | + | | | | + | | + | |
| Chronometric Tasks—MT | | | | | | | | + |
| | | | | | | | | + |
| | | | | | | | | + |
| | | | | | | | | + |
| | | | | | | | | + |
| | | | | | | | | + |

different group factors depending on whether the ECT makes a demand on memory (either STM or LTM), labeled ECT+Mem, as in the Sternberg and Posner paradigms, or makes no memory demand (labeled ECT:RT), as in the Hick and OMO paradigms.

In Carroll's analysis, the average g loading of the PTs=.35; of the ECT:RT, g loadings =.43. The group factors had the following average loadings: PT-Verbal=.61, PT-Spatial= .47; ECT:RT=.42, ECT+Mem:RT=.58, and ECT:MT=.51.

The results of Carroll's (1991b) definitive analysis of the Kranzler–Jensen (1991) data help to explain why the simple correlations between single psychometric and single chronometric measures are typically so relatively small. It is because a large part of the variance in each type of variable, both psychometric and chronometric, consists of a group factor + the variable's relatively large specificity, which the variables do not have in common. It appears at this stage of analysis that there may be only one factor (here called g) that is common to both the psychometric and chronometric domains. Other studies, too, have found that the degree of correlation between RT measures and various PTs depends on the size of the tests' g loadings (Hemmelgarn & Kehle, 1984; Smith & Stanley, 1983). No significant or systematic relationship has been found between RT and any other psychometric factors independent of g.

A quite similar result to that of Carroll's analysis is indicated by the correlations (Miller & Vernon, 1996, from Tables 6, 8, and 10) based on eight diverse RT tests (with mean RTs ranging between 800 and 1800 ms) and 10 subtests of the Wechsler Preschool and Primary Scale of Intelligence-Revised (WPPSI-R) given to 109 children of ages 4 to 6 years (mean IQ 107, SD 13). On the full 18×18 matrix of correlations among the RT and WPPSI variables, I have performed a nested factor analysis, in which the general factor is extracted first, followed by extraction of the remaining significant factors. There is a large general factor (*g*), accounting for 36 percent of the total variance on which all of the 18 variables have substantial loadings; and there are only two group factors (with eigenvalues >1), accounting for 12 and 6 percent of the total variance. These two factors represent the well-established Verbal and Nonverbal Performance factors found in all of the Wechsler test batteries. The RT tests' significant loadings are entirely confined to the *g* factor, with the single exception of 2-choice RT (for Colors: same or different), which was loaded on both *g* and the Performance factor. After the *g* factor was extracted, the RT variables yielded no coherent group factor(s). There was nothing at all that looked like an RT factor independent of *g*. Whatever non-*g* variance remained in the whole battery of eight RT tests was specific to each test. In the Kranzler–Jensen data set factor analyzed by Carroll (1991), as described above, there were distinct RT group factors, independent of *g*, representing RT tasks that either did or did not make demands on memory, whether STM or LTM. In the Miller and Vernon study, however, none of the RT tasks involved either STM or LTM. The study by Miller and Vernon also included five separate tests of STM, which, besides having large *g* loadings, share the same group factor with the WPSSI subtests, but the memory tests have relatively weak correlations (−.20 to −.40) with RT. The interaction of memory and processing speed seem to be fundamental to *g*.

If different sets of PTs that are based on different contents, such as verbal (V), quantitative (Q), and spatial (S), are factor analyzed along with parallel sets of chronometric tests (ECTs) based on the same contents, will the three types of content be represented by three group factors, V, Q, S, or will the group factors represent only the distinction between PT and ECT? This question was answered in a study by Levine, Preddy, and Thorndike (1987), based on samples of school children in the 4th, 7th, and 10th grades. The result of a hierarchical factor analysis is predictable from the previously described studies. All of the PTs and ECTs loaded about equally on the *g* factor (average loadings for PTs = .43, for ECTs = −.41). There were three group factors, V, Q, S, on which only the PTs showed significant loadings; and there was a single RT factor, independent of *g*, with loadings unrelated to the test contents. The specific content features of the ECTs were not represented in the factor structure. The lone RT factor represents a source of variance unique to the ECTs regardless of content; it is probably a psychomotor factor common to all of the six of the ECTs, which were administered with the same apparatus and procedure. Again, we see that the correlation between nonspeeded psychometric scores and RT is mediated indirectly by the relationship of each type of variable to a higher order factor (*g* in this case). But the stark conclusion that RT is a better measure of *g* than it is of any psychometric group factors, if it even reflects any of them at all, should be held in abeyance until this possibility is confirmed by further studies.

Although both PTs and ECTs, when factor analyzed together in one correlation matrix, have substantial loadings on the general factor of the whole matrix, the question arises whether the resulting general factor is a different *g* or might in some way be a better *g* than that extracted exclusively from a psychometric battery, such as the Wechsler Intelligence

Scales. To answer the question as to a "better" *g*, one must ask "better for what?" Hence, an external criterion is needed to assess which of the two *g* factors is a better predictor of the criterion.

Exactly this question was posed and researched in two studies using various types of factor analysis and structural equation modeling (another statistical technique for identifying latent variables and their interrelationships) (Luo & Petrill, 1999; Luo, Thompson, & Detterman, 2003). The subjects were 568 elementary school children. The psychometric battery was the 11 subtests of the Wechsler Intelligence Scale for Children-Revised (WISC-R); the chronometric battery was RTs on each of six ECTs; and the criterion variable for scholastic performance was the Metropolitan Achievement Test (MAT), assessing reading, math, and language.

The data in the exceptionally instructive analyses by Luo and Petrill (1999) and Luo et al. (2003, 2006) are like the studies previously reviewed, with the typical results depicted in Table 9.2. That is, both the psychometric and the chronometric measures are loaded on a single large common factor, *g*, and they have no lower-order group factors in common — the group factors are specific to each domain. More importantly, the position of the *g* axis of the combined psychometric and chronometric tests *vis-à-vis* the axis of the general factor of scholastic achievement is very close to the *g* axis of the combined correlation matrix. The *g* of the combined matrix is, in fact, a slightly better predictor of scholastic achievement than the psychometric battery alone (validity coefficients of about 0.5 and 0.6, respectively). So *g*, whether psychometric or chronometric, is pretty much one and the same general factor. Because there is virtually no general or scholastic information content in the simple chronometric ECTs, the authors inferred that speed of information processing as measured by the ECTs is an important mechanism mediating the observed correlations between the general factor of the PTs (WISC-R) and the general factor of scholastic achievement (MAT). They also submit that information processing speed may not be the only elemental cognitive component mediating the correlation between *g* and scholastic performance. Certain nonchronometric variables, particularly memory processes, contribute to the correlation between psychometric *g* and scholastic performance independently of processing speed. As hypothesized later on, there appears to be a synergistic interaction between processing speed and memory processes that accounts for most, or perhaps all, of the *g* variance.

### Failed Explanations of the RT–IQ Correlation

The revival of research on the RT–IQ relationship engendered numerous criticisms of both the idea of there being a correlation between RT and IQ and the actual findings showing such a correlation. Critics were seldom previously involved in this field of research and apparently had no intentions of collecting appropriate data to empirically check their speculations. The seeming aim was not to encourage further research but rather to "explain away" the RT–IQ correlation in various superficial ways that, if accepted as valid, would dampen interest in pursuing this line of investigation. After all, to many psychologists at that time it seemed too implausible that individual differences in anything as apparently mindless and trivial as a person's RT performance could be causally related to individual differences in something so miraculously complex as human intelligence.

The total extant evidence at that time, however, forced the debate to shift from contesting the existence of the RT–IQ correlation itself to its interpretation. The arguments hinged mainly on the *direction* of causation. Is the RT–IQ correlation explained by *bottom–up* processing (i.e., faster RT → higher IQ), or by *top–down* processing (i.e., higher IQ → faster RT)? The top–down view was favored by critics of the essential findings on the RT–IQ correlation. They argued that higher-level cognitive processes are the causal factor in the acquisition and performance of lower-level skills, such as RT performance, and this therefore explains the observed IQ → RT relationship. According to this view, mental speed *per se* plays no causal role but only reflects the consequences of higher-level processing. The IQ → RT correlation is claimed to come about because individuals with higher intelligence learn new skills more readily, take greater advantage of subtle cues in the task format, discover more optimal performance strategies, and the like. Hence they are winners in the RT game. This is all quite plausible, however, uninteresting. For if we accept the *top–down* theory to explain the IQ → RT correlation, there is nothing new to be learned about the nature of individual variation in IQ from further studies of this phenomenon. It would be a blind alley for intelligence researchers wishing to go beyond conventional psychometrics and descriptive statistics.

The *bottom–up* theory, on the other hand, proposes the opposite direction of causality in the RT → IQ connection. It aims to search out the root causes of intelligence, or *g*, with speed of cognitive processing as a crucial construct and the measurement of RT and IT essential tools. The program's research rationale is essentially this: Individuals with faster speed of information processing (hence faster RT) thereby take in more information from the environment per unit of exposure time — attending to, processing, and consolidating a larger proportion of the information content into LTM, later to be retrieved as knowledge and cognitive skills, thus causing learning at a faster rate, and developing all the variety of "higher-cognitive processes" identified as intelligence. The highest-order common factor in assessments of all these cognitive aspects of individual differences is *g*. Thus, individual differences in speed of processing are posited as a fundamental, or bottom–up, cause of *g*. To discover the physical basis of processing speed requires that cognitive neuroscientists must search further down the causal chain for the precise mechanisms linking the brain to RT and other behavioral expressions of *g*. In this analytic-reductionist program, experiments require the precise quantitative data afforded by chronometry, on the one hand, and by direct imaging and measurement of brain variables, on the other, each domain acting alternately as the independent or the dependent variable in experimental designs.

Probably the most comprehensive critique of the bottom–up formulation is that of Longstreth (1984), who suggested a number of possible artifacts in RT methods that might account for the IQ–RT correlations. Even the linear slope of the Hick function and its relation to IQ were attributed to the specific order of presenting the 0–3 bits RT tasks. Some of Longstreth's complaints could be contradicted with empirical evidence available at that time (Jensen & Vernon, 1986). Later on, the experimental artifacts and confounds Longstreth and others held responsible for the RT–IQ findings were empirically investigated in independent studies designed to control or manipulate the possible effects of each claimed artifact. In most of these studies, the artifacts could not account for the IQ–RT correlations or other theoretically relevant features of the RT data, and in some studies, eliminating one or more of the imputed artifacts even increased the RT–IQ correlation

(Kranzler, Whang, & Jensen, 1988; Neubauer, 1991; Neubauer & Fruedenthaler, 1994; Larson & Saccuzzo, 1986; Larson, Saccuzzo, & Brown, 1994; Smith & Carew, 1987; Vernon & Kantor, 1986; Widaman & Carlson, 1987; for reviews: Deary, 2000a, pp. 156–160; Jensen, 1998a, pp. 238–248).

Rather than detailing each of these often-complex studies, it would be useful here to simply list the most prominent of the various failed hypotheses proposed as explanations of the RT–IQ correlation.

**Motivation**   Higher-IQ subjects are assumed to be more motivated to excel on any cognitive task and hence show faster RT. However, direct indicators of variation in degree of autonomic arousal and effort (e.g., pupillary dilation) show that on tasks at any given level of difficulty higher-IQ subjects register *less* effort than lower-IQ subjects. The authors of a definitive study of this hypothesis concluded "**. . .** more intelligent individuals do not solve a tractable cognitive problem by bringing increased activation, 'mental energy' or 'mental effort' to bear. On the contrary, these individuals show less task-induced activation in solving a problem of a given level of difficulty. This suggests that individuals differing in intelligence must also differ in the efficiency of those brain processes which mediate the particular cognitive task" (Ahern & Beatty, 1979, p. 1292).

**Practice and learning**   The idea is that higher-IQ subjects learn the task requirements faster and benefit more from practice and therefore excel at RT tasks. Considering the simplicity of the ECTs on which RT is measured, this seems a most unlikely explanation for the RT–IQ correlation observed in samples of very bright college students performing tasks that are understood even by preschoolers and the mentally retarded. But more telling is the fact that the RT performance of lower-IQ individuals does not gradually catch up to that of higher-IQ individuals, even after many thousands of practice trials spread over many sessions. The idea that RT differences on the simplest ECTs are the result of differences in ability to grasp the task requirements is not at all borne out by any evidence. Beyond the first few practice trials, the rank order of individuals' RTs is maintained throughout extended RT practice up to an asymptotic level. Although prolonged practice results in a slight overall improvement in RT, its magnitude is exceeded by individual differences in RT.

**Visual attention, retinal displacement, and eye movements**   This explanation assigns the causal locus to peripheral rather than central processes. It applies to choice RT tasks in which the choices vary in the spatial location of the RS, which varies randomly from trial-to-trial. Task difficulty and RT vary with the number of RS alternatives, as in the Hick task. As the number of alternatives increases, the subject's visual focus of attention will, on average, be further removed from the randomly presented RS. The RT on the more difficult task conditions, therefore, could be a function of the degree "retinal displacement" of the RS and the slight automatic eye movement needed to bring the RS into acute focus (i.e., foveal vision). RT is slightly faster for stimuli focused on the fovea than on other retinal areas (peripheral vision). Thus the increase in RT as the number of RS alternatives is increased might not be a measure of cognitive processing speed as a function of information load, but only a peripheral ocular effect. This explanation, however, is contradicted by the finding that the increase in RT as a function of the number of alternatives in the RS occurs even when the RS alternatives (e.g., different colors) are always presented in exactly one and the same location, on which the subject's vision is directly focused immediately before the appearance of each RS, assuring acute foveal vision on virtually all trials. The result shows that the number of RS alternatives per se (i.e., the ECT's cognitive information load), rather than variation in the visual location of the RS, is the main cause of a subject's RT to the RS. Individual variation in RT is clearly a central, not a peripheral, phenomenon.

**Speed-accuracy trade-off and other strategies**   This is the idea that subjects adopt a particular strategy for maximizing their performance, and smarter subjects can figure out more effective strategies. An obvious strategy is to increase speed at the expense of increasing response errors. When subjects are instructed to use this strategy, it effectively decreases RT and increases error rate. So it was assumed that higher-IQ subjects are more likely to discover this RT/error trade-off strategy, which would explain the negative RT–IQ correlation. The fallacy in this explanation, however, is that the trade-off effect, which can be experimentally manipulated by instructions, turns out to be entirely a within-subjects effect, not a between-subjects effect. Therefore, it cannot be causally related to individual differences in IQ — an entirely between-subjects source of variance. If the brighter subjects in any sample had faster RTs because they are more likely to spontaneously adopt a strategy of speed/accuracy trade-off, then we should expect (a) a positive correlation between IQ and errors and (b) a negative correlation between RT and errors. But what is actually found is the opposite direction of both of these correlations. The brighter subjects have both fewer errors and faster (i.e., shorter) RT. Other strategies also have been experimentally investigated with respect to various ECTs, and there is no evidence that the IQ–RT correlation on ECTs can be in the least explained by individual differences in the use of particular strategies. In probably the most comprehensive examination of the strategies hypothesis, the investigators concluded "Alternatives to strategy theories of g must be pursued if progress is to be made in explaining general intelligence" (Alderton & Larson, 1994, p. 74).

**Speeded psychometric tests**   One of the most frequently proclaimed explanations for the RT–IQ correlation is that both the RT task and the correlated PT share the same common speed factor when all or part of the PT is speeded. This idea is totally invalidated by clear-cut evidence. PTs show at least as high or higher correlations with RT when they are given without time limits as when the same tests are speeded. Since the test-speed notion was the most prevalent explanation of the RT–IQ correlation when I began researching RT some 30 years ago, from the beginning it has been routine practice to administer every PT without time limit and instruct subjects to attempt every item and take as much time as they wish. To prevent subjects' awareness of the time others take on the test (to prevent a "keeping up with the Joneses" effect), every subject is tested alone in a quiet room. After giving instructions, the examiner leaves the room. If the subject's performance is timed, it is without the subject's knowledge. So studies of the RT–IQ relationship conducted in our chronometric laboratory are not confounded by test-taking speed. Individual differences in ad libitum time on PTs are not significantly correlated with RT but with a personality factor — extraversion—introversion (assessed by the Eysenck Personality Inventory in our early studies of RT). Though extraversion is negatively correlated (about −.40) with total ad libitum testing time, it is correlated neither with test scores nor with RT.

### Special Chronometric Phenomena Related to Psychometric Abilities

**Intraindividual variability (RTSD)**    This is the trial-to-trial fluctuation in RT, measured as the SD of a subject's RTs over a given number of trials ($n$), henceforth labeled RTSD. It has not yet been established whether RTSD represents a truly random *within*-subject fluctuation or is a regular periodic phenomenon that is disguised as a random variation because of the asynchrony of response *vis-à-vis* the randomly timed appearance of the RS. This is a crucial question for future research. If the apparently random fluctuation in RTs represented by RTSD is not just an artifact of such asynchrony but is truly random, it would suggest a kind of "neural noise," for which reliable individual differences in magnitude might explain $g$. It would be even more important theoretically if it were established that RTSD reflects a true neural periodicity, or oscillation in reaction potential. What has been well established so far is that variation in RTSD is a reliable individual differences phenomenon and, more importantly, individual differences in RTSD are correlated with $g$ and IQ at least as much, and probably more, than is the mean or median RT or any other parameters derived from RT data (Jensen, 1992; Baumeister, 1998). When RTSD shows a lower correlation with IQ than does RT, it is usually because of the much lower reliability of RTSD. The disattenuated correlation of RT and RTSD with IQ generally favors RTSD. So this is one of the most central issues for RT research. Chaper 11 fully explains the importance of intraindividual variability for a theory of the RT–$g$ correlation.

A provocative study based on 242 college students found that RTSD measured in various chronometric tasks is positively correlated with a psychometric scale of neuroticism, suggesting that trait neuroticism is a reflection of "noise" in the brain's neural control system (Robinson & Tamir, 2005). It is puzzling, however, that mean RT itself showed no significant correlation with neuroticism, given the intrinsically high correlation between RT and RTSD found in most studies. This calls for further investigation to be entirely convincing as a broadly generalizable empirical fact. It could turn out to be merely a scale artifact.

In fact, the mean RT and RTSD are so highly correlated as to seem virtually redundant in a statistical and factor analytic sense, much as would be the redundancy of diameters and circumferences of various-sized circles. The near-perfect correlation between RT and RTSD reflects the fact that across individuals there is a virtually *constant proportionality* between mean RT and RTSD, as measured by the coefficient of variation (CV = SDRT/RT). Yet, RT and RTSD behave quite differently in a number of ways, including their relation to the RT task's information load, as shown in Figure 9.11, based on 1400 subjects. The Hick task's mean RTs are plotted as a function of bits, and the means of RTSD are plotted as a function of the actual number of response alternatives for the same data. The difference between these functions is a real phenomenon, not a mathematically necessary artifact. RT increases as a linear function of bits (i.e., $\log_2$ of the number of S–R response alternatives), while RTSD increases linearly as a direct function of the number of response alternatives.

As explained previously (Chapter 4, pp.), another statistic, the *mean square successive difference* (MSSD) or $\sqrt{\text{MSSD}}$, should supplement RTSD in future research on intraindividual variability. I have not found MSSD ever being used in RT research, unfortunately. Because RTSD comprises not only random fluctuation between trials but also any systematic or nonrandom trends in the subject's performance, such as practice effects, it may confound quite different sources of RT variation. MSSD, however, measures purely the absolute

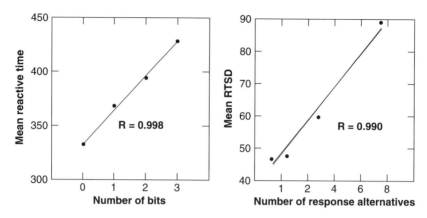

Figure 9.11: Mean RT and Mean RTSD, based on 1400 subjects, plotted as a function of bits (for RT) and number of response alternatives (for RTSD). (From data in Jensen, 1987a.)

RT fluctuations between adjacent trials and therefore is uncontaminated by possible systematic trends in the subject's performance. It would be a major empirical achievement to demonstrate that the magnitude of MSSD can be manipulated experimentally, as it could provide the means for testing the hypothesis that the observed random fluctuations in a subject's trial-to-trial RTs is essentially the result of the timing of the RS being randomly out of synchrony with the subject's regular oscillation of reaction potential. This is a challenge for the techniques of experimental psychology. If behavioral evidence of a regular oscillation is found, the next question obviously is whether individual differences in the rate of oscillation in reaction potential, as an operationally defined construct, has a physical basis in brain functioning. Its investigation would call for the techniques of neurophysiology.

**The "worst performance rule"**  This is another phenomenon that must be explained by any theory of the RT–IQ correlation. Though it has been known for decades, it received little attention. It was not until 1990 that it was given a name — the *worst performance rule* (WPR) — in connection with an excellent large-scale study that definitively substantiated the WPR (Larson & Alderton, 1990). This surprising phenomenon is the fact that, in Larson and Alderton's words, "The worst RT trials reveal more about intelligence than do other portions of the RT distribution." The WPR, tested with quite different RT tasks in a college sample, was further established by Kranzler (1992). One study based on a very heterogeneous sample (ages 18 to 88 years) did not show the WPR, perhaps because of the study's many differences from previous studies, in the RT tasks, tests, and procedures used (Salthouse, 1998). Coyle (2003) found the WPR to apply in children in the average range of ability (mean IQ=109) but not in gifted children (mean IQ=140), and related this finding to Spearman's "law of diminishing returns," which states that conventional IQ tests are less *g* loaded for individuals at higher levels of ability, because a larger proportion of their variance in abilities is invested in various, more specialized, group factors, such as verbal, numerical, and spatial abilities, consequently leaving a proportionally smaller investment in *g* (Jensen, 2003). This suggests that WPR phenomenon depends mainly on the *g* factor rather than on a mixture of abilities including their non-*g* components.

The WPR is demonstrated with RT tasks as follows: across *n* test trials each subject's RTs are rank ordered from the fastest to the slowest RT. (To minimize outliers often the two most extreme RTs are discarded, so RTs on *n*−2 trials are ranked). The RT–IQ correlations *within* each rank then are calculated for the entire subject sample. Consistent with the WPR, it is found that the resulting within-rank correlations systematically increase from the fastest to the slowest RT. It is also important for a theory of the RT–IQ correlation to note that the within-ranks coefficients of variation (CV=SD/mean) are perfectly correlated (*r*=.998 in the Larson and Alderton study) with the within-ranks RT–IQ correlations. This close connection between the WPR phenomenon and the RTSD–IQ correlation implies that if intraindividual fluctuation in RT across trials (i.e., RTSD) is considered the more "basic" phenomenon, then the WPR is simply a necessary derivative consequence of the RTSD–IQ correlation, which, as proposed in Chapter 11, is the chief causal mechanism in explaining the RT–IQ correlation and possibly the basis of *g* itself.

The WPR can also be displayed graphically by comparing two groups that differ in IQ. Note that the differences between the group means increase going from the fastest to the slowest RTs, as shown in Figure 9.12 based on simple RT measured with the same apparatus and procedures in both groups.

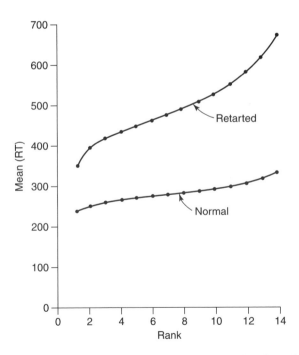

Figure 9.12: Mean simple RT plotted as a function of rank order (from fastest to slowest) of each individual's RTs, for groups of young adults with normal intelligence (mean IQ 120) and with mental retardation (mean IQ 70). (From Jensen, 1982a, p. 291, with permission of Springer.)

**Convertability between RT and response errors**   In studies of the RT–IQ relation, little investigative attention has been paid to response errors, that is, making the wrong response when there are two or more possible choices. The intense focus on purely mental speed variables in most RT–IQ studies has resulted in a neglect of the role of errors in the RT–IQ correlation. But there is also the fact that experimenters have usually tried to minimize error rates by keeping the RT tasks simple and by stressing accuracy as well as speed in the preliminary instructions. In some studies, only RTs for correct responses are used in the data analysis, and though errors may be automatically counted, they are seldom reported. Moreover, unless the RT task is fairly complex and the number of test trials is quite large, individual differences in the very small error rates are almost like single-digit random numbers and have near-zero reliability. Hence at present most possible generalizations about errors are relatively impressionistic and call for more systematic empirical verification.

However, there are two phenomena that have been rather informally but consistently observed in many of the studies in the Berkeley chronometric laboratory. The first is that *response errors increase with task complexity* (as indicated by mean RT). The second is that as task complexity increases, up to a point, the increasing errors are increasingly correlated with IQ; but beyond that critical level of task complexity, errors are more correlated *with* IQ than is RT. As tasks increase in complexity, there is a trade-off between RT and errors in their degree of correlation with IQ.

This suggests that the difficulty level of cognitive tasks when ranked in terms of RT should be similar to that of the same tasks as measured by their error rates, assuming that RTs and error rates both had enough variability to allow their reliable measurement in the same data set. This hypothesis was explicitly tested in a study that compared a group of university students with grade-school children (ages 8 to 9) on the Semantic Verification Test (SVT), previously described on page 20 (Jensen et al., 1988). It was necessary to have two groups of widely differing ability levels in order to obtain relatively error-free speed measures in one group and to get a large enough number of errors in the other group to ensure a sufficiently wide range of variation in the item's error rates under untimed conditions. There were 14-item types each given (with different permutations of the letters) seven times in a random order, totaling 84 items in all. The university students were given the SVT as a RT task, emphasizing both speed and accuracy in the instructions. (Their mean RTs on each of the 14-item types ranged from about 600 to 1400 msec, with an average error rate of 7 percent. The mean RTs were correlated $-.45$ with scores on the untimed Advanced Raven Matrices, which is as high as the correlation between the Raven and the Wechsler IQ (WAIS) in university samples.) The very same SVT items were given to the school children in the form of an *untimed* paper-and-pencil test of 28 items (two items for each of the 14 SVT item types). The instructions made no mention of speed and urged that every item be attempted. The required response was to circle either YES or NO that accompanied each SVT item in the test booklet. The children's average error rate was 18 percent. The essential finding: the children's mean errors on the 14 SVT items had a rank correlation of .79 (.90 when corrected for attenuation) with the mean RTs of the corresponding SVT items in the university sample. In other words, the more difficult the item was for the children, the greater was its mean RT for the university students. This high congruence between item RTs and item error rates suggests the possibility of using very high-IQ college students' mean RTs on easy test items as a means for obtaining objective ratio-scale measures of item difficulty in untimed PTs for children. This phenomenon of the "conversion" of RTs to response error rates (or conversely to *p*-values) calls for further study. Aside from its possibly practical applications, it seems a crucial point for a theory of the RT–IQ correlation and hence for a theory of *g* itself.

**Task complexity and the RT–IQ correlation**   It has long seemed paradoxical that RT has low to moderate correlations with IQ, the correlation increasing as a function of task complexity (or length of RT) while the time taken per item on conventional PTs is much less correlated with total score (or with IQ on another test) based on the number of items being scored as correct. The times taken per Raven matrices item, for example, show near-zero correlations with RT. The true-score variance of test scores depends almost entirely on the number right (or conversely, the number of error responses). The relationship between RT and task complexity or cognitive load of the items to be responded to (i.e., the RS) has been a subject of frequent discussion and dispute in the RT–IQ literature (e.g., Larson & Saccuzzo, 1989). I have examined virtually the entire literature on this seeming paradox, but rather than giving a detailed account of all these empirical studies, I will simply summarize the main conclusions that clearly emerge from a wide range of studies. These findings can be illustrated by a couple of studies that were specifically directed at analyzing the relationship of task complexity to the RT–IQ correlation.

But first, a few words about the definitions of *complexity* in this context. One or another of five clear operational criteria of task complexity is generally used: (1) the average of the subjective ratings made by *N* judges of *various RT tasks'* "complexity"; (2) the amount of uncertainty as an increasing function of the number of choices (response alternatives) that are associated with the *n* different RS, such as the difference between SRT and CRTs based on two or more stimulus-response alternatives; (3) the theoretically presumed number of distinct mental operations that are required for a correct response, such as the difference between adding two digits and adding three digits; (4) the difference between (a) *single tasks* that make a minimal demand on memory and (b) *dual tasks* requiring that one item of information RS1 be held in memory while performing the interposed task RS2–RT2, then performing RT1; and (5) various tasks' mean RTs used as a measure of complexity. All of the above conditions except 1 and 5 can be experimentally manipulated as independent variables while RT is the dependent variable.

Subjective judgment (condition 1) is probably the most questionable measure, although, as predicted by the Spearman–Brown formula, the mean ranking of tasks for "complexity" would gain in validity by aggregating the rankings by an increasing number of judges. A study of the SVT (described on page ) in which a group of 25 college students were asked to rank the 14-item types of the SVTs for "complexity" showed that subjective judgments of item complexity do have a fair degree of objective validity (Paul, 1984). The raters were naïve concerning the SVT and its use in RT research. The mean ratings on "complexity" of the 14 SVT items (from least complex = 1 to most complex = 14) had a rank-order correlation of +.61 with the items' mean RTs obtained in another group of students (*N* = 50).

The hypothesized relationship of the RT–IQ correlation to task complexity is shown in Figure 9.13. The level of complexity at the peak of the curve is not constant for groups of different ability levels. Although the relative levels of complexity on different tasks can be ranked with fair consistency, the absolute complexity level varies across different ability levels. The peak of the curve in Figure 9.13 occurs at a shorter RT for adults than for children and for high IQ than for low IQ groups of the same age. The peak level of task complexity for correlation with IQ in college students, for example, is marked by a mean RT of about 1 s; and for elementary school children it is between 2 and 3 s. But there has not been enough systematic parametric research on this point to permit statements that go beyond these tentative generalizations.

A direct test of the hypothesis depicted in Figure 9.13 was based on eight novel speed-of-processing tasks devised to systematically differ in difficulty or complexity (Lindley, Wilson, Smith, & Bathurst, 1995). They were administered to a total of 195 undergraduate college students. IQ was measured by the Wonderlic Personnel Test. The results are summarized in Figure 9.14. This study affords a clue to what is probably the major cause

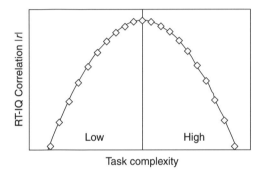

Figure 9.13: The generalized relationship of the RT–IQ correlation to task complexity. The absolute value of the correlation coefficient |r| is represented here for graphic clarity, although the empirical RT–IQ correlation is always a negative value, with the very rare exceptions being attributable to sampling error.

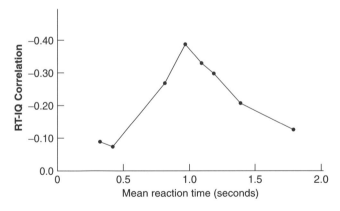

Figure 9.14: The RT–IQ correlation plotted as a function of mean RT for 105 undergraduates. (From data in Lindley et al., 1995.)

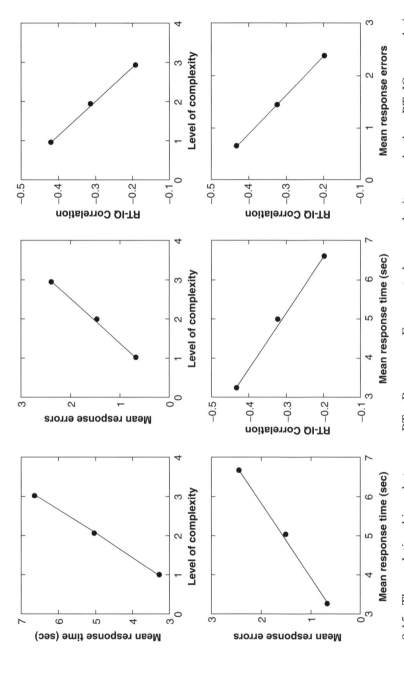

Figure 9.15: The relationships between RT, Response Errors, task complexity, and the RT–IQ correlation for homogeneous RT tasks at three levels of experimentally controlled task complexity. (From data in Schweizer, 1998.)

of the very wide range of RT–IQ correlations reported in various studies. The correlation is influenced by two conditions: (1) test complexity and (2) the mean and range of IQ in the subject sample, as the peak of the complexity function shifts to longer RTs as the mean IQ declines. Therefore, the significant RT–IQ correlations fall within a relatively narrow range of task complexity for various groups selected from different regions of the whole spectrum of ability in the population. Hence, when it comes to measuring general intelligence by means of RT there is probably no possibility of finding any single RT task with a level of task complexity that is optimally applicable to different samples that ranges widely in ability. The average RT–IQ correlation in the general population on any single task, therefore, represents an average of mostly suboptimal complexity levels (hence lower RT–IQ correlations) for most of the ability strata within in the whole population.

The optimum level of task complexity for the IQ–RT correlation is hypothesized to occur near the RT threshold between error-free responses and error responses. This is the point on the complexity continuum beyond which RT becomes less correlated (negatively) with IQ and errors become increasingly correlated (negatively) with IQ.

This hypothesis of a trade-off between RT and errors in the RT–IQ correlation and the Errors–IQ correlation was tested in a study expressly designed for this purpose (Schweizer, 1998). In order to study the relationships between errors, RT, and the RT–IQ correlation, the RTs and the number of errors had to be measured entirely on the High side of the complexity function shown in Figure 9.13, resulting in mean RTs ranging between 3 and 7 s; and even then the error rates averaged only 16 percent. Three sets of different RT tasks were used (numbers ordering, figures ordering, mental arithmetic). In each set, the task complexity was experimentally controlled as the independent variable to produce three distinct levels of complexity, determined by the number of homogeneous mental operations required to make a correct response. IQ was measured as the averaged scores on the Wechsler test (WAIS-R); subjects were 76 university students (mean IQ=120.4, SD 9.6).

Figure 9.15 shows the results (averaged over the three different RT tasks) for the hypothesized functional relationships between the key variables. The consistent linearity of the relationships shows that it is possible to devise cognitive tasks that vary unidimensionally in complexity.

Unfortunately, a graph of the relation between complexity and the Error–IQ correlation is not possible with the given data. The Error–IQ correlations were said to be very small and only the two largest of the nine possible correlations were reported, both significant ($-.24$ and $-.28$, each at $p < .05$); but they evinced no systematic relationship to task complexity. It would probably require a considerably greater range of complexity and error rate to adequately test the relation between task complexity and the Errors–IQ correlation. In typical PTs it is so problematic to measure *item complexity* that the term is usually used synonymously with *item difficulty*, measured as the error rate (or percent passing) when all item responses are scored as either right or wrong. Then, of course, the relationship between item difficulty and the Error–IQ correlation is a foregone conclusion. The correlation between item *response times* and IQ based on right–wrong scoring is typically very low, but this is mainly because there are so many different causes of error responses to test items, except in item sets that have been specially constructed to differ in difficulty along some unitary dimension of complexity. The meaning of complexity in chronometric tasks is discussed further in Chapter 11.

# Chapter 10

# Sensory Intake Speed and Inspection Time

Although tests of RT appear to be very simple compared to the typical items in psychometric tests, it turns out that reaction times, even simple reaction time (SRT), are actually complex phenomena with various points of entry for the play of individual differences in sensory, motor, and cognitive components. These can be empirically analyzed. Table 10.1 lists these pivotal points.

[a]SRT or CRT = total time elapsed between Stage 2 (RS onset) and 2Bc (overt response). The premotor and motor stages of RT are known to be heterogeneous with respect to individual differences; i.e., they show different correlations with age, IQ, and other external variables. For example, the premotor stage but not the motor stage of RT is correlated with IQ. Here we are not referring to *movement time* (MT) as this term is used in connection with RT devices that involve the subject's release of a home button and pressing a response button. The motor aspect of interest here involves only the release of the home button *per se*. It is the intrinsic motor aspect of the initial overt response (pressing or releasing) the home button, i.e., the RT (also referred to as decision time, or DT) when the response console has both a home button and a response button.

The *motor stage* actually begins well before the initial overt response, as shown by an ingenious experiment on SRT and CRT in a group of school-age children who were prenatally exposed to alcohol (ALC) as compared with a normal control (NC) group (Simmons, Wass, Thomas, & Riley, 2002). Of interest here is only the partitioning of the RT into its *premotor* and *motor* components, referred to as the premotor RT (or PMRT) and the motor RT (or MRT). This division of the total RT was accomplished by obtaining electromyographic (EMG) recordings from electrodes attached to the subject's arm muscle while the subject was performing the RT task. A typical result for a single subject on SRT is illustrated in Figure 10.1, in which A indicates the stimulus onset (RS), B the beginning of the motor response, and C the overt response. The time interval A–B is PMRT, and the interval B–C is MRT. For children aged 8–9, the average PMRT is about 300 ms for SRT and about 400 ms for CRT; the corresponding MRTs are about 70 and 80 ms, so PMRT is about four to five times greater than the MRT. The PMRT is more sensitive than MRT by age and IQ differences. (Both variables showed significant deleterious effects of prenatal exposure to alcohol, with a larger effect for CRT than for SRT.)

An important conclusion from this demonstration is that even SRT is not causally unitary, but is a composite of at least two distinct variables, PMRT and MRT. Along with sensory transduction of the visual stimulus (RS) from the retina to the visual cortex (about 50 ms), the total time for perceptual intake and analysis of the RS apparently includes about 80 percent of the total RT, slightly more for CRT than for SRT. Moreover, the percentage of RT constituting the total PMRT increases as a function of RS complexity. Shortened exposure or weaker intensity of the RS requires longer analysis and results in a slower RT. In terms of signal detection theory, a longer-lasting stimulus better overcomes the level of random background "noise," thus increasing discrimination of the RS and hence the speed

Table 10.1: Sequential analysis of sensory, cognitive, and motor components of SRT and choice reaction time (CRT).

| Stage | Subject's activity |
|---|---|
| 0. Pretask | (a) Instructed on nature and demands of task |
| | (b) Practice trials; familiarization |
| Task proper | |
| 1. Preparatory signal | (a) Vigilance, expectancy, focused attention |
| 2. Response stimulus (RS) onset[a] | |
|     A. Premotor stage | (a) Sensory transduction of RS to brain |
| | (b) Stimulus apprehension (in SRT) |
| | (b') plus discrimination (in CRT) |
| | (c) Response selection (in CRT) |
|     B. Motor stage | (a) Efferent nerve propagation |
| | (b) Recruitment of motor response |
| | (c) Overt response execution |

and probability of a correct response to the RS, particularly as the RS increases in complexity, as in multiple CRT tasks. The background "noise" consists both of distracting external task-irrelevant stimuli and internal random neural activity. Thus an important question is: how long a stimulus exposure at a given level of stimulus intensity is required to accurately discriminate a given difference between two stimuli, absent any motor component?

## Inspection Time (Visual)

The first systematic attempt to answer this question, conceived as a problem in psychophysics is known as the *inspection-time paradigm*, originally developed by Vickers, Nettelbeck, and Willson (1972). This paradigm attracted little interest outside psychophysics until it was discovered that measures of individual differences in inspection time (IT) had remarkably substantial correlations with IQ (Nettelbeck & Lally, 1976). Since then a large literature on IT, based on a great many empirical studies and theoretical discussions, has grown up mainly in differential psychology. As there are now comprehensive reviews and meta-analyses of all the research on IT, it would be pointless to present here more than a summary of the main findings and issues, noting the most critical questions that are still in need of further empirical studies. To virtually enter the entire literature on IT, with detailed summaries and discussions of many specific studies, readers should consult the key references (Brand & Deary, 1982; Deary, 1996, 2000a, Chapter 7; Deary & Stough, 1996; Luciano et al., 2005; Nettelbeck, 1987, pp. 295–346, 2003, pp. 77–91).

    The best-known method for measuring IT is described in Chapter 2, pp. 33–35. The test stimulus and backward mask are presented tachistoscopically by means of light emitting diodes (LEDs). The IT paradigm differs most importantly from all other chronometric

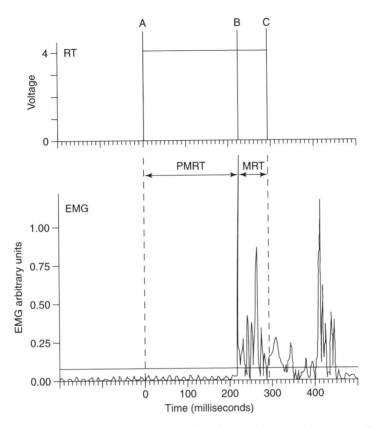

Figure 10.1: Simple RT of a 9-year-old, showing the time intervals for the PMRT (A–B) and MRT (B–C) and the corresponding electromyograph showing the muscle activation preceding the overt RT by 80 ms. (From Simmons et al., 2002, with permission of the Research Society on Alcoholism.)

paradigms by completely eliminating any demand for a speeded motor response. Measurements of IT, therefore, reflect no motor component whatsoever. The sensory discrimination called for in the IT paradigm is so simple that every individual would perform with 100 percent accuracy if there were no severe time constraint on the presentation of the stimulus. For example, following a preparatory stimulus to focus the subject's attention, the subject must decide which one of two parallel lines presented side by side is shorter, where the shorter line is only half the length of the longer line (see Figure 2.14, p. 33). The two lines are exposed tachistoscopically and are followed after the very short IT exposure interval by a "backward mask," which completely covers and obliterates the test stimulus, thereby allowing no further inspection. The subject is encouraged to take as much time as needed to make a decision. Across multiple trials the interval between the test stimulus and the mask is systematically varied until the experimenter has zeroed-in on the stimulus duration for which the subject can discriminate accurately on, say, 75 percent of the trials, i.e., the midpoint between 50 percent chance of guessing and 100 percent accurate discrimination. Figure 10.2 shows the performance of a single subject whose IT is about 12 ms by the 75 percent criterion.

The originator of the IT paradigm, the psychophysicist Douglas Vickers, first conceived of IT as a theoretical construct: namely, as the minimal amount of time needed for a single *inspection*, defined conceptually as the minimal sample of sensory input required for the simplest possible discrimination. The particular discriminated stimuli, consisting of two vertical parallel lines of conspicuously unequal lengths, was devised as the operational estimate of visual IT (VIT). Because somewhat different discriminanda of similar simplicity produced different values of IT and some subjects' performance was affected by extraneous perceived cues (e.g., an apparent motion effect on the occurrence of the masking stimulus) critics of the IT paradigm argued that IT probably involves something more than sheer speed of sensory intake. Today most researchers, including Vickers, generally agree with the position stated by Mackenzie, Molley, Martin, Lovegrove, and McNicol (1991) that "IT now appears, not as the time required for initial stimulus intake, but as the time (or some consistent portion of the time) required to solve a problem: to make the required discrimination or matching judgment about the stimulus items" (p. 42).

Studies of RT and studies of IT have been quite insular and, until recently, *both* paradigms have seldom entered into one and the same study. Also, the comparatively large amount of time subjects take to make a judgment following the IT discriminandum has been virtually ignored. This IT response interval clearly does not meet any definition of RT, because it is intrinsic to the IT paradigm that there be no explicit time constraint on the subject's response. However, anyone who has taken the IT test, or observed others taking it, will have noticed that as the duration of the IT interval between the test stimulus and

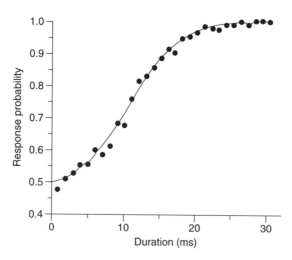

Figure 10.2: The probability of a correct discrimination as a function of stimulus duration for an individual. Each data point represents the mean of a number of trials. The individual's IT is usually estimated as the average stimulus duration with .75 probability of correct responses. (From Deary & Stough, 1996, with permission of the American Psychological Association.)

the backward mask is made shorter, the overt DT increases. On the more difficult discriminations, the subject must think longer about what was seen in order to increase the probability of making the correct decision. To what extent does the subjects' trying to "read back" or recover the quickly fading memory trace of the stimulus play a part in IT performance? The fact that a decision is involved in IT implies that some cognitive effort is involved, as well as an involuntary sensory speed component. Although IT is only a fraction of the length of SRT or CRT, it is not necessarily any less complex than RT psychologically. The empirically least questionable difference between RT and IT is the absence of a motor component in IT. The discovery of other possible differences between RT and IT will depend on knowing more about the difference between the construct validity of each paradigm, their disattenuated correlation with each other, and differences in their correlations with external psychometric variables.

## *Reliability of IT*

It is important to determine the reliability of IT for the specific subject sample and the particular apparatus and procedure used in the study. A proper estimate of reliability is essential for the interpretation of the correlation of the IT measurements with any external variables or for establishing its true construct validity.

The reliability for any given IT procedure is positively related to the number of test trials and the variance of IT in the study sample. However, it has not yet been established, as it has for RT, how closely the reliability of IT is predictable from the Spearman–Brown formula relating reliability to the number of items (or trials) in the test. Such parametric information is useful for any quantitative science. Investigators must also take note of the difference between the two main types of reliability coefficient: internal consistency and test–retest. They are conceptually different and each is appropriate for a given purpose. The fact that they typically have quite similar values is just an empirical coincidence rather than a conceptual or mathematical necessary. There is no rational basis for inferring one from the other. Internal consistency reliability (split-half or preferably coefficient alpha, which is the average of all possible split-half reliability coefficients for the given data set), is an index of the factor homogeneity of the measurements. Test–retest reliability estimates the temporal stability of the mean (or median) measurements obtained on different occasions separated by a specified time interval. It needs to be taken into account in evaluating replications of a study. If the test–retest stability coefficient is low, the results of attempted replications will vary more erratically than if the stability coefficient is high, depending on the number of subjects.

Fortunately, there is a full meta-analysis of the reliability of IT based on virtually all of the IT studies available prior to 2001 that compared reliability coefficients of IT and IQ: a grand total of 92 studies comprising 4197 subjects (Grudnik & Kranzler, 2001, Table 2), shown here in Table 10.2. In those studies that reported both internal consistency and test–retest reliability, only the coefficient with a lower value was used in calculating the mean reliability, therefore resulting in a conservative estimate of the average reliability, weighted by sample size. For the total of all samples the weighted mean reliability is .805. The mean for adults is .815; for children .782. Correction of the sample means for restriction of range increases the reliability about +.04 overall. These analyses indicate that the

Table 10.2: Mean and variance of reliability coefficients for IT and IQ.[a]

| Meta-analysis | IT | | IQ | |
| --- | --- | --- | --- | --- |
| | *M* | *V* | *M* | *V* |
| Total | .805 | .015 | .948 | .033 |
| Adults | .815 | .015 | .942 | .031 |
| Children | .782 | .013 | .945 | .046 |
| IT task type | | | | |
| Visual | .833 | .008 | .942 | .031 |
| Auditory | .815 | .015 | .942 | .031 |
| Strategy users/nonusers | | | | |
| Users | .805 | .015 | .948 | .033 |
| Nonusers | .815 | .015 | .942 | .033 |

[a]Based on Table 2 (p. 527) in Grudnik and Kranzler (2001).

reliability of both visual and auditory IT (AIT) is nearly as high as the typical psychometric tests used for individual diagnosis and could easily be made higher simply by increasing the number of test trials. The mean reliability of the IQ tests based on their standardization samples is about +.10 to +.15 higher than the mean reliability of IT as obtained in these studies. This is quite remarkable, in fact, because the IQ tests are composed of multiple diverse items that allow chance variation and subtest specificities to "average-out," whereas IT is an extremely homogeneous task administered in relatively few trials compared to the number of items in a typical IQ test. With more extensive testing, the reliability of IT could well exceed that of the best IQ tests.

### Construct Validity of IT

This is the least researched and least satisfactorily answered question regarding IT. The construct measured by IT is still in question. The only way its construct validity can be determined is by means of correlational analyses, such as factor analysis and structural equation modeling, in the company of other well-defined variables, and by establishing its convergent and divergent validity in comparison with other chronometric methods.

Yet we can draw a few tentative conclusions from the relevant correlations now reported. IT is clearly different from RT, because when entered in a multiple regression equation along with RT it makes an independent contribution to the correlation with IQ or psychometric *g*. Also, IT shows a different profile of correlations with various psychometric tests from that of RT, being more highly correlated with visual-spatial tests than with verbal tests than it is with RT. And IT tends to be less strongly *g* loaded than RT. A second-order general speed factor, however, links IT to the third-order factor of psychometric *g* and hence to general IQ tests (O'Connor & Burns, 2003). A striking feature of O'Connor's and Burns's factor analysis of several different speed-of-processing tasks, including IT, the Hick paradigm, the odd-man-out (OMO) paradigm, and other speeded tasks, is that the Hick RT variables have near-zero loadings (average loading = + 0.007) on the first-order factor (labeled Visualization Speed) on which IT has its largest loading (+ 0.350), but the OMO RT has its largest loading (+ 0.469) on this same Visualization factor. IT is loaded −0.021 on the DT factor on which the Hick RT variables (for 2-, 4-, and 8-response alternatives) have loadings of + 0.788, + 0.863, and + 0.773, respectively. (All the MT variables are loaded entirely on a separate factor. Thus it appears that IT may not even be a component of either 2-, 4-, and 8-CRT in the Hick paradigm, none of which demands much visual discrimination, but IT and OMO share a factor that demands a higher degree of visual discrimination. Hence it is certain that IT and RT are not factorially equivalent or interchangeable variables.

Is "speed of sensory intake" a suitable construct label for IT? To qualify for the breadth expected of a construct, the variable of VIT would have to be shown to correlate highly with other measures of sensory intake, such as its procedural counterpart, AIT. Strangely, those who research VIT have worked so independently of those who research AIT that the two paradigms have not yet been directly compared empirically in any systematic fashion. Both VIT and AIT, however, are quite comparably correlated with IQ, as shown in Table 10.3.

However, VIT and AIT could possibly have different correlations with different subtests or with different group factors or specificity in the various IQ tests. The latent trait represented by three distinct auditory discrimination tests, including AIT, have correlations with each other in the range .60 to .75 and their general factor is amazingly correlated .64 with psychometric $g$ as represented by the general factor of three very highly $g$-loaded psychometric tests (Deary, 2000, p. 217, Figure 7.9). Although both VIT and AIT are similarly correlated with $g$, they have different specificities ($s$) and probably one or more unshared group factors ($F$) related specifically to auditory ($a$) or visual ($v$) processes. The simplest

Table 10.3: Meta-analysis of the IT × IQ correlations (raw and corrected[a]) reported in published studies (from Grudnik & Kranzler, 2001, with permission from Elsevier).

| Meta-analysis | $N$ | $K$[b] | Raw | | Corrected[a] | |
|---|---|---|---|---|---|---|
| | | | Mean | Variance | Mean | Variance |
| Participants | | | | | | |
| Total | 4197 | 92 | −.30 | .03 | −.51 | .00 |
| Adults | 2887 | 62 | −.31 | .03 | −.51 | .00 |
| Children | 1310 | 30 | −.29 | .03 | −.44 | .01 |
| IT task type | | | | | | |
| Visual | 2356 | 50 | −.32 | .03 | −.49 | .02 |
| Auditory | 390 | 10 | −.30 | .04 | −.58 | .07 |
| Strategy users/nonusers | | | | | | |
| Users | 205 | 9 | −.34 | .03 | −.60 | .00 |
| Nonusers | 160 | 9 | −.46 | .03 | −.77 | .00 |

[a]Correlations corrected for the sampling artifacts of sampling error, attenuation, and range variation.
[b]Number of independent studies. $N$ is the number of individuals.

likely factor models for VIT and AIT, for example, would be VIT$=g+Fv+s$; and AIT$=g+Fa+s$. (*Note*: by definition, the specificities, $s$, of VIT and AIT are different and uncorrelated.)

A whole area of investigation is opened by questions about the construct validity of IT and of how it is related to RT and other chronometric variables. So far, the main interest in IT has been its quite remarkable correlation with IQ. Few other indicators of IT's external or ecological validity have yet been sought.

## Correlation of IT with IQ

Probably, the most consistent and indisputable fact about IT is its substantial correlation with IQ. There are three comprehensive reviews and meta-analyses of the findings (Nettelbeck, 1987; Kranzler & Jensen, 1989; Grudnik & Kranzler, 2001). Since the most recent meta-analysis, by Grudnik and Kranzler, is cumulative over all the published data, its conclusions, as far as they go, are definitive. The IT $\times$ IQ correlations are shown in Table 10.3. The correlations were also corrected for artifacts due to sampling error, attenuation, and range variation. The results are shown here separately for adults and children, visual and AIT, and for groups claiming to use a conscious strategy (such as an apparent motion effect) versus those claiming no such strategy in their performance of the IT task. The overall corrected mean $r$ is $-.51$ ($-.30$ prior to correction). An important observation is that there is no significant variation in the corrected correlations across the various studies. Also, the correlation differences between children and adults, and between visual and AIT are statistically nonsignificant. The difference between strategy users ($-.60$) and nonusers ($-.77$) appears substantial, but because of the small number of samples ($K = 9$) it was not tested for significance. Grudnik and Kranzler (2001, p. 528) also point out those children who are diagnosed as reading disabled perform very differently on IT than do children of normal reading ability. These two groups show the same *magnitude* of RT $\times$ IQ correlation, but with *opposite* signs: disabled $r = +.44$, nondisabled $r = -.44$! This finding suggests that IT (and possibly the degree of discrepancy between visual and AIT) might be useful in the differential diagnosis and prescription for treatment of various reading disabilities.

## IT in a Psychometric Factor Space

Although it is amply proven that the speed of IT is substantially correlated with IQ ($r \approx .50$), it cannot be assumed that the IT is mainly correlated with $g$, even though IQ and $g$ are highly correlated. This is because the IQ is not a unidimensional variable but is an inconsistent amalgam of a number of different factors and test specificity, such that $g$ seldom accounts for more than 50 percent of the total common factor variance among the various subtests in any given test battery. This allows the possibility that $g$ might contribute less to the IQ $\times$ IT correlation than one or more of the other factors. This possibility has been looked at in several studies, with mostly ambiguous or inconsistent results, probably because of differences in the study samples, differences in the psychometric test batteries, and differences in the IT apparatuses and procedures.

Consider a battery composed of 11 psychometric tests: Advanced Raven and Multidimensional Aptitude Battery (MAB) and 36 chronometric variables based on 7 different elementary cognitive tasks (ECTs), first published in a study by Kranzler and Jensen (1991). In an orthogonalized hierarchical factor analysis (Schmid–Leiman) of these data, Carroll (1991b) found that IT was loaded only .19 on the highest-order factor (at the second-order), which he labeled "Decision Speed and Psychometric *g*." For comparison, the average of the loadings of SRT and CRT (Hick paradigm) on this factor are .37 and .56, respectively, and the OMO RT is loaded .62. The loadings of the 10 MAB psychometric subtests on the same general factor range from .28 to .46, with an average loading of .37. The Raven Advanced Progressive Matrices loaded .41. Apparently, IT has much less common variance with the psychometric tests than does the RT variables. IT has its largest loading on a first-order factor that Carroll labeled "Decision speed — Hick and IT tasks" — a first-order common factor of Hick RT and IT residualized from the second-order general factor. Even the first order factor of "crystallized intelligence" (defined by MAB subtests: Vocabulary, Information, Comprehension, Similarities, Picture Completion) have loadings on the second-order *g* averaging .30 and ranging from .28 to .35. The largest Pearson *r* that IT has with any one of the six independent RT measures in this study is only .216 (Hick 8-CRT). Yet IT has similar correlations with two of the MAB subtests (Picture Completion .313, Spatial .221).

Another way to examine the relationship of IT to psychometric *g* is by the method of correlated vectors (fully described in Jensen, 1998b, Appendix B). In this case, the method consists essentially of obtaining the rank-order correlation between the column vector of disattenuated *g* loadings of the subtests of an intelligence battery with the corresponding column vector of each of the various subtests' disattenuated Pearson correlations with speed of IT. When this method has been applied to RT measures, there is typically a high *positive* correlation between the vector of the subtests' *g* loadings and the vector of the subtests' correlation with speed of processing in RT tasks.

The correlated vectors method was applied to IT by Deary and Crawford (1998), using the 11 subtests of the Wechsler Intelligence Scale for Adults-Revised (WAIS-R) to obtain the subtests' *g* loadings. In three independent representative subject samples, the *g* and IT vectors were *negatively* correlated (−.41, −.85, −.31) — an opposite result than is usual for RT (and various biological measures, such as brain size and cortical evoked potentials). Evidently, IT gains its correlation with IQ through factors other than *g*, such as the *g*-residualized common factor of the WAIS-R Performance subtests (see Deary, 1993). But could this result be peculiar to the WAIS-R, in which the so-called Verbal subtests typically have somewhat larger *g* loadings than the so-called Performance subtests? If the theoretical or "true" *g* is conceived to be independent of the verbal–nonverbal distinction, and is not intrinsically more (or less) related to verbal than to nonverbal tests, this finding by Deary and Crawford could be just a peculiar artifact of the WAIS-R battery. Its generality would depend on replications of the same outcome in different test batteries. It might be that verbal tests, which tap possibly the highest cognitive functions to evolve in the human brain, are intrinsically more sensitive than nonverbal tests to reflecting *g* differences — a question that can be answered only empirically.

A seeming contradictory result is found in a battery of 14 diverse subtests, including both verbal and nonverbal subtests as well as 4-CRT and IT. Nettelbeck and Rabbitt (1992) found that both RT and IT have their largest loadings (−.69 and −.71) on the second-order hierarchical *g* factor. The main difference between IT and CRT is that CRT has its only salient loading on g, while IT has a salient loading (−.33) also on a first-order factor which has salient loading also on the WAIS subtests of Picture Arrangement and Block Designs (with g loadings of .69 and .82). The method of correlated vectors applied to this test battery used by Nettelbeck and Rabbitt shows a correlation of +.95 between the vector of subtests' g loadings and the vector of subtests' correlations with speed of IT — a result strongly opposite to the finding of Deary and Crawford (1998) described above. No definitive explanation of the huge discrepancy between the results in these two studies can be given, except to note the two main differences between the studies: (1) In the Deary and Crawford study the largest g loadings in the battery were on the verbal tests, whereas in the Nettelbeck and Rabbitt study the largest g loadings were on the nonverbal performance tests; and (2) the subject samples differ considerably in age and ability level. The D&C sample was fairly representative of the general adult population, whereas the N&R sample was relatively elderly and of high IQ, with ages ranging from 54 to 85, and a mean IQ at the 90th percentile of the general population. In any case, these conflicting studies are puzzling and leave the present issue unresolved. Factor analyses based on larger batteries than just the 11 subtests of the Wechsler batteries should help in locating IT in the psychometric factor space, as far as that might be possible.

At present, the safest conclusion seems to be that VIT is related to most conventional psychometric batteries both through g and through one or more lower-order factors representing aspects of visual processing speed (or auditory processing speed in the case of AIT). As is also true for any measure of RT, much of the variance in IT, is unrelated to traditional measures of psychometric intelligence (see Chaiken, 1994). The hypothesized relationships of VIT and AIT to each other and to psychometric g are shown in terms of a Venn diagram in Figure 10.3, in which the area of each circle represents the total standardized variance

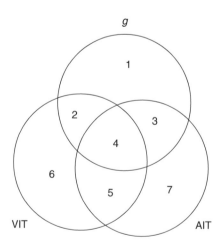

Figure 10.3: Venn diagram showing hypothetical relationships between Visual and Auditory ITs and psychometric g. Areas: 1+2+3+4 = total variance of psychometric g,  2+4=g variance of VIT, 3+4=g variance of AIT, 4+5 = common variance of VIT and AIT, 5 = first-order common factor variance of 6 = specific variance of VIT. 7 = specific variance of AIT.

($\sigma^2=1$) of each of the variables (psychometric $g$, VIT, AIT). The areas of overlap represent latent variables, i.e., proportions of variance common to any two or three of the depicted variables, two of which (VIT and AIT) are directly measured, whereas $g$ is wholly a latent variable common to all cognitive abilities.

It should be remembered, however, that the external validity and importance of IT does not depend on the number or nature of the factors on which it has its salient loadings in any kind of factor analysis of conventional psychometric batteries, which scarcely encompass the entire array of cognitive abilities. There is a risk in validating new measures of intelligence against older established measures, because these may fail to reflect certain abilities or traits that are important predictors of performance in life. In most factor analyses of psychometric batteries, what appears as the large *specificity* of IT could have important correlates in certain cognitive aspects of living that are untapped by most psychometric tests which would constitute the external validity of IT. A striking example of this is seen in a study by Stough and Bates (2004). A comprehensive test of secondary school scholastic achievement was correlated with a battery of four standard psychometric tests (correlations in parentheses): Raven Advanced Progressive Matrices (.23), Verbal IQ (.39), Figural-Spatial IQ (.39), and IT (−.74). The multiple $R$ of the three standard tests with Scholastic Achievement is .44. Including IT along with the three standard tests raises the multiple $R$ to .79, hence the *incremental validity* of IT is .67. Also, in a principal components analysis of all five measures, which yields two significant components (i.e., eigenvalues > 1), IT had the largest loading of any of the variables on the first component (−.75). (The second principal component has significant loadings only on IT and Achievement.) This remarkable finding, based on $N = 50$, of course, calls for replication.

The latest addition to the puzzlement so far regarding the nature of the correlation between IT and IQ is a most impressive large-scale behavior genetics study based on 2012 genetically related subjects in Holland and Australia (Luciano et al., 2005). This study used latent trait models to help decide between the "bottom up" versus the "top down" hypotheses of whether individual differences in perceptual speed influence differences in intelligence (*IQ*) or differences in IQ cause differences in perceptual speed. The correlation (about −.40) between IQ and IT as a latent trait did not fit either the bottom-up or the top-down causal models but is best explained as the result of pleiotropic genes affecting variation in both IT and IQ, without implying that either variable has any causal effect on the other. (The genetic phenomenon known as *pleiotropy* is explained in Chapter 7, pp. 127–129) The authors concluded: "This finding of a common genetic factor provides a better target for identifying genes involved in cognition than genes which are unique to specific traits" (p. 1).

# Chapter 11

# Theory of the Correlation Between Response Time and Intelligence

At present we can make no claims of a single unified theory that explains the correlation between response time (RT) and intelligence in neurophysiological terms. The purpose of this chapter, therefore, is to point out the critical features of the chronometric evidence to date that must be considered in formulating a theory of the RT–IQ correlation.

We can no longer regard seriously the earlier criticisms of attempts to explain the basis for this remarkable correlation by questioning its existence or validity. The RT–IQ correlation *per se* is an empirical fact as thoroughly proved as any phenomenon in the behavioral sciences. Yet some cognitive theorists still seem to regard this correlation skeptically. If they acknowledge it at all, they trivialize it by pointing out that some animals — frogs, pigeons, and cats, for example — show quicker RT under certain species-related stimulus conditions than humans, yet it would be ridiculous to regard these species as more intelligent than humans. This *non sequitur* ignores differences in species-specific behavior, speed of sensory transduction, brain circuitry, nerve conduction velocity, and the neural distances traversed, not to mention the striking differences in the apparatus and procedures required for testing RT in different species.

Also, it should be kept in mind that human individual differences in highly complex decision-making tasks are a poor reflection of basic speed of information processing, because individuals also differ in the amounts of information they take into account in solving a complex problem. Individuals also differ in the number of distinct elements in a problem they have encountered previously and thereby differ in the benefit of positive transfer. The relatively simple ECTs used in chronometric research undoubtedly allow more experimental control of task complexity and reduce unknown sources of variance that enter into individual differences in the speed of decision making in complex psychometric tests or in "real-life" problems. A chess master playing against a duffer, for example, makes much faster decisions than when playing against another master, because master-level chess is played at a much more complex level of strategy and most of the players' moves call for a much greater amount of information processing. Decision speed accounts for over 80 percent of the variance in ratings of chess skill even among expert tournament players (Burns, 2004).

A useful theory serves as scaffolding for the discovery of new facts about a given realm of phenomena. A heuristic theory should have quite general predictive and explanatory power across all of the RT and IT paradigms discussed in previous chapters. In formulating a general theory, one must "read through" the experimental and sampling "noise," task specificity, or other unique phenomena associated with any particular study.

First, we should specify the essential variables whose intercorrelations the theory intends to explain.

Response time : The *primary variables*, RT (and inspection time (IT)), are measured in milliseconds by a chronometer under laboratory conditions in response to some form of elementary cognitive task (ECT) in which the reaction stimulus (RS) is either visual or auditory.

The theory does not encompass movement time (MT), except to note that MT fundamentally differs from RT. Although RT and MT are both conceptually and factor analytically distinct variables, in practice MT usually contaminates RT to some extent, which typically reduces the RT–IQ correlation. This is especially true when the response console does not separate the measurements of RT and MT, and it is also true, though to a much lesser degree, even when RT and MT are measured separately by using a console on which the subject releases a home button (RT) and presses a response button (MT). Such MT effects on various kinds of choice RT (CRT) can be mostly gotten rid of by subtracting or partialling out simple reaction time (SRT) from the CRTs elicited by ECTs of greater complexity than the paradigm for SRT (Jensen & Reed, 1990). The theory also takes into account *secondary variables* derived from the basic RTs on each test trial over *n* trials: the *mean* RT (RTm), *median* RT (RTmd), and the *standard deviation* of RT (RTSD) over *n* trials. Other derived measures are the *slope* of RT as a function of variation in task complexity and the *mean square successive difference* (MSSD), a measure of intraindividual variability over *n* successive trials that does not reflect systematic trends in the absolute level of RT, such as a practice effect (see Chapter 4, p. 67).

*Psychometric g:* The present theory is not concerned with the correlation between RT and scores on any particular psychometric test except insofar as it is loaded on the *g* factor. The theory concerns the correlation of RT with the *g* factor ideally extracted from an indeterminably large battery of nonspeeded tests representing a great variety of mental abilities. The theoretical, mathematical, and empirical status of the *g* factor has been spelled out elsewhere (Jensen, 1998b; 2002), but it may be useful to summarize the main points.

- The number of *specific* cognitive abilities is indeterminably large. *Cognitive* refers to conscious voluntary activity involving stimulus apprehension, discrimination, decision, choice, and the retention of experience or memory.
- Individual differences in any specific cognitive ability have many causes: neurological limitations on basic information processing; knowledge or skills acquired through interactions with the environment; and opportunity, predisposition, and motivation for particular kinds of experience.
- Individual differences in many abilities can be assessed with psychometric tests. Individual differences in all cognitive abilities are positively correlated to some degree, indicating they all have a common source of variance. A mathematical algorithm can be used to analyze the matrix of correlations among many diverse ability measurements to reveal the significant independent common factors in the matrix, termed principal components or factors. About 50 such independent factors have now been reliably identified. However, they differ greatly in generality and importance in life.
- The factors can be visualized as a triangular hierarchy, numbering about 40 of the least general primary factors at the first (lowest) level, eight or nine more general second-order factors at the next higher level, and the one most general factor (*g*) at the apex. Each factor in this hierarchical structure represents a statistically independent component of individual differences. The total variance of any particular test comprises *g* and possibly one or more of these first- and second-order factors. (The total variance on any test also includes *specificity*, an unidentified but reliable source of variance that is specific to a

particular test, and random *measurement error* or unreliability.) At present, these are all the reliable factors revealed by factor analyses of hundreds of various tests of human abilities. Some future psychometric or chronometric tests might result in the discovery of additional factors that are independent of all the presently identified factors, except *g*.

- Every cognitive ability that shows individual differences is loaded on the *g* factor. Tests differ in their *g* loadings, but their *g* loadings are not related to any particular kinds of knowledge or skills assessed by various tests. The possible cognitive indicators of *g* are of unlimited diversity. Although *g* is certainly not the *only* important factor, its extraordinary generality makes it the *most* important factor. In a large battery of diverse cognitive tests, *g* typically accounts for some 30–50 percent of the total population variance in test scores, far exceeding any of the subordinate factors.

- It is also important to understand what *g* is *not*. It is not a mixture or an average of a number of diverse tests representing different abilities. Rather, it is a *distillate*, reflecting the single factor that all different manifestations of cognition have in common. In fact, *g* is not really an ability at all. It does not reflect the tests' contents *per se*, or any particular skill or type of performance. It defies description in purely psychological terms. Actually, it reflects some physical properties of the brain, as yet unknown, that cause diverse forms of cognitive activity to be positively correlated, not only in psychometric tests but in all of life's mental demands. IQ scores are an attempt to estimate *g* and they typically do so quite well. The average correlation between various IQ scores and *g* is about .85. But because an IQ test is necessarily just a vehicle for *g*, it inevitably reflects other broad factors as well such as verbal, numerical, and spatial abilities, besides the specific properties of any particular IQ test. Yet, under proper conditions, the IQ is a good estimate of an individual's relative standing on *g*.

- Although *g* is manifested to some degree in every expression of cognition, some tasks and abilities reflect *g* much more than others. Usually, *g* is positively related to differences in the *complexity* of a task's cognitive demands. Also, *g* is the platform for the effective use of the abilities represented at the level of first-order factors and for the expression of musical and artistic talents.

- More than any other factors, *g* is correlated with a great many important variables in the practical world, like educability, job proficiency, occupational level, creativity, spouse selection, health status, longevity, accident rates, delinquency, and crime.

- Also, *g* is uniquely correlated with variables outside the realm of psychometrics, particularly biological variables having behavioral correlates:
  - The *heritability* (i.e., the proportion of genetic variance) of various tests is directly related to the tests' *g* loadings.
  - *Inbreeding depression of test scores* is a genetic effect that lowers a quantitative trait. It results from the greater frequency of double-recessive alleles in the offspring of genetically related parents, such as cousins. The degree of inbreeding depression on various mental test scores is strongly related to the tests' *g* loadings. The larger the *g* loading, the greater is the effect of inbreeding depression on the test scores.
  - Certain anatomical and physiological brain variables are related to differences in tests' *g* loadings: brain size, brain glucose metabolic rate, the latency and amplitude of cortical-evoked potentials, brain nerve conduction velocity, brain intracellular pH level, and certain biochemical neurotransmitters. Thus, *g* reflects biological components of

intelligence more strongly than any other psychometric factors (or any combination thereof) that are statistically independent of *g*.

The group factors residualized from *g* have peculiarly low or nonexistent correlations with all of these biological variables, with the exception of the visual–spatial and probably the auditory factors. It seems likely that the variance represented by all of the other group factors results from the experiential interaction of *g* with various forms of content afforded by the environment, such as specific verbal and numerical knowledge and skills. Tests of vocabulary, verbal analogies, and word similarities, for example, are all so very highly *g* loaded as to leave almost no reliable *non-g* variance to be independently correlated with genetic or other biological variables. A higher level of *g* allows greater acquisition and retention of verbal knowledge in a social and scholastic environment. In contrast, tests of spatial and auditory abilities retain fairly substantial factor loadings on their respective group factors independently of *g*. Although variance in spatial and auditory skills probably comes about to some degree through interaction with certain visual and auditory aspects of environmental stimulation, it appears to have a more substantial biological anlage, independent from that of *g*, than is the case for verbal and numerical knowledge.

## The Two Basic Chronometric Variables: RT and RTSD

For any given individual, the RT on any single trial is a virtually worthless measure, regardless of the sensory modality or the intensity of the RS. A reliable measure of an individual's RT requires that a considerable number of RT trials be summarized by their central tendency, either the *mean* (RTm) or the *median* (RTmd). Their reliability coefficients, and hence their potential for correlation with any other variables, depends on the measurement technique used and the number ($n$) of practice trials that precede the $n$ test trials on which RTm or RTmd are based.

Because the distribution of individual RTs is nonsymmetrical, being always skewed to the right, RTm is necessarily always a larger value than RTmd. (In the normal distribution or any perfectly symmetrical distribution the mean and median are identical.)

Intraindividual trial-to-trial variability in RT, measured as the standard deviation of individual RTs over $n$ trials (RTSD) is more important from a theoretical viewpoint than the measures of central tendency. The important fact for our theory is that, given a large number of trials, there is a near-perfect correlation between individual differences in RTm and RTSD. Hence the coefficient of variation (CV = RTSD/RTm) for individuals is virtually a constant value for any given RT task. Empirically measured diameters and circumferences of different-size circles are no more highly correlated than are RTm and RTSD. The slight deviations of their correlation coefficient from unity are simply measurement errors. In the case of the correlation between RTm and RTSD, measurement error consists of some contamination by systematic trends in RTs across trials (e.g., a practice effect) in addition to the inevitably imperfect reliability of both RTm and RTSD when the RTs are measured in some limited number of trials. It is also important for theory that, although the reliability of RTmd is generally slightly higher than that of RTm, RTmd is typically less correlated with RTSD than is RTm.

## Oscillation Theory

Despite the near-perfect correlation between RTm and RTSD, it can be postulated that RTSD is theoretically the more basic variable than RTm. The theoretical argument for this supposition, then, is the point at which we must introduce the oscillation theory of individual differences in RT. Its basic empirically observed elements are the following:

**A.** Every individual has a naturally imposed *physiological lower limit* of RT. This limit increases systematically as a function of increasing task complexity. For any individual, the limit varies slightly from day to day and throughout each day, reflecting subtle changes in physiological states associated with diurnal fluctuations in body temperature, time elapsed since the ingestion of food, certain drugs, previous amounts of sleep, and emotional state (mood, anxiety level, and the like).

**B.** Despite these diurnal fluctuations in physiological states, there are highly reliable individual differences in the natural physiological limit of RT.

**C.** Reliable individual differences in the physiological lower limit of RT are of *lesser magnitude* than individual differences in any externally evoked RTs that exceed the lower physiological limit of RT.

**D.** In any succession of single evoked RTs over *n* trials, the intraindividual trial-to-trial fluctuations of RT necessarily vary in magnitude only uni-directionally, i.e., *above* the individual's present physiological limit of RT.

**E.** Trial-to-trial variation in the RTs of a given individual appears random across *n* successive trials, assuming there is no systematic practice effect throughout the *n* trials. Under this condition, the RTSD is a valid measure of intraindividual trial-to-trial variation in RT.

**F.** Any single RT comprises the sum of the individual's physiological limit *plus* a positive deviation from that limit. In a given test session of *n* trials, an individual's RTm is the sum of RT/*n*.

**G.** The total distribution of a given individual's RTs over *n* trials is not symmetrical but is necessarily *skewed to the right* (i.e., longer RT), because there is a lower physiological limit of RT and there is no naturally imposed upper limit of RT.

**H.** Because of the positive skew of the RT distribution, an individual's RTm will necessarily exceed the RTmd (i.e., RTm > RTmd). Also, RTm is more highly correlated with RTSD than is RTmd. In fact, RTm and RTSD are both aspects of intraindividual variability, conceptually distinct but perfectly correlated, like diameters and circumferences of various circles. The perfect correlation between individual differences in RTm and RTSD is not a mathematical necessity or a statistical or measurement artifact, but represents a natural phenomenon.

**I.** Theoretically, individual differences in RTm can be causally derived from whatever causal phenomenon is measured by individual differences in RTSD, but the reverse is not the case. Therefore, intraindividual variability in RTs (measured as RTSD) represents a theoretically more fundamental aspect of individual differences in speed of information processing than does RTm.

**J.** The hypothesized cause of RTSD is the oscillation of neural excitatory potential.

### Ancillary Comments on the above Propositions E and G

**E'.** Trial-to trial fluctuations in RT as measured by RTSD constitute two potential sources of variance: (1) systematic changes in RT resulting from practice effects spread across a series of trials, and (2) nonsystematic or purely "random" variation in RTs. "Random" is in quotes, because, although the RTs may look random and conform to statistical tests of randomness, this apparent randomness may be attributable to the virtual randomness of the experimenter-controlled times of presentation of the RS on each trial.

A measure of fluctuations in RTs that cannot reflect practice effects or any other systematic trends affecting the RTs across trials is the MSSD previously explicated (Chapter 4, p. 67). In the absence of any systematic trends, however, RTSD and MSSD are perfectly correlated. A fair number of practice trials lessens systematic trends in RTs in the test trials. Practice effects are also lessened by a higher level of stimulus–response compatibility in the given ECT. Practice effects are more pronounced and RTs are faster when there is an optimum interval between the preparatory signal and the RS. Practice effects are also more evident when the subject has some voluntary control over the timing of the occurrence of the RS, such as by pushing the home button to elicit the RS, which appears after a brief fixed interval (and to a lesser degree for two or three different fixed intervals that occur at random). Presumably, over some *n* trials individuals gradually develop a subjective moment-to-moment feeling of fluctuation in their readiness to make a quick response and this feeling becomes more synchronized with the voluntary elicitation of the RS. The subjective sense of "readiness" and the practice effects on the quickness of RTs are also enhanced by the subject's receiving informative feedback of the RT on each successive trial.

But it is still empirically unknown precisely which testing procedures result in the most informative measures of RTSD (or MSSD) with respect to their correlations with *g*. This can only be determined by parametric studies of a given ECT.

**G'.** Theoretically, measures of individual differences in the degree of positive skewness (Sk) of the RT distribution are perfectly correlated with individual differences in both RTm and RTSD. However, it is technically impossible to show reliable measures of individual differences in Sk. The hindrance to achieving a suitable reliability coefficient for measures of Sk, such as we can obtain for RTm and RTSD, is inherent in the indices of Sk, which are based on the difference between an individual's RTm and RTmd or on the ratio of the third and second moments of the individual's distribution of RTs. Such measures are so plagued by the worst disadvantages of difference scores as to be condemned to near-zero reliability. Because of this, an index of Sk is unsuitable as a measure of individual differences. If Sk happens to show a significant correlation with any other variable the correlation should be suspected as being some kind of artifact.[1]

### The Worst Performance Rule

This phenomenon was described in Chapter 9 and Figure 9.13, p. 185. An individual's slower RTs, which constitute the positively skewed tail of the individual's RT distribution, are more indicative of the level of IQ or *g* than are the individual's faster RTs. Also, the RT variance *between* subjects increases monotonically from the fastest to the slowest RTs produced by each subject. These observations are a consequence of the differing degrees of Sk of the

RT distribution for individuals, which is predictable from the theory of individual differences in neural oscillation rate.

It is also important to note that the effect of increasing task complexity is to increase the slower RTs much more than the faster RTs, and the increasing task complexity also magnifies the RT differences *between* groups that differ in g (e.g., normal and retarded groups) much more for the slower than for the faster RTs. Increasing task complexity has a constant multiplicative effect on all of a subject's RTs. This effect is reflected in the RTm and RTSD for any given subject and also for the differences between subjects. These RT effects are evident in all ECT paradigms that differ in task complexity, or information processing load. So general are these effects not only in studies of individual differences in RT, but also in studies of learning and memory or gains in task proficiency with increasing practice (except when performance is limited by a low ceiling effect), that I have called them the first and second laws of individual differences:

*First Law of Individual Differences.* Individual differences in learning and performance increase monotonically as a function of increasing task complexity.

*Second Law of Individual Differences.* Individual differences in learning and proficiency increase with continuing practice and experience, provided there is not an intrinsically low ceiling on task performance.

### *Task Complexity*

This concept, which is prominent in comparing various RT paradigms, does not have a univocal definition or a single objective unit of measurement independently of RT. It may even be premature to impose a too restrictive criterion for its scientifically useful meaning. In the present context, the term *complexity* refers to the information load of the given task, which is usually described in terms of (1) the number of task elements the subject must attend to; (2) the degree of prior uncertainty about the required response; (3) the degree of essential stimulus or response discrimination involved in the task; (4) the degree of stimulus–response compatibility; (5) the number of decisions that have to be made to achieve a correct response; or (6) the amount of prior-learned information that must be retrieved from short- or long-term memory to make the correct response. Reliable measures of RT are highly sensitive to remarkably slight variations in any of these conditions of complexity. In groups of college students, for example, there is an average difference of 40–50 ms even between SRT and two-CRT, when there is an optimal preparatory signal and the RS in each case is simply a bright green light going "on," always in the same location (SRT), or randomly in one of two locations (CRT).

# Neural Oscillation as the Basis of Intraindividual Variability

Although neural oscillation is proposed as a hypothetical construct in the present theory, it has well-established analogues in neurophysiology that are an empirical reality, generically called "brain waves." These periodic variations in the brain's electrical potentials are usually measured in microvolts from various electrodes attached to the scalp; the weak potentials are amplified and recorded by the techniques of electroencephalography. The periodicity of the electrical potentials between different regions of the brain registers the spontaneously synchronized effect among a multitude of neurons (Abeles, 2004; Buzsaki & Draguhn, 2004; Ikegaya et al., 2004; Strogatz, 2003; Bodizs et al., 2005). The effective action of the brain in cognition and behavior depends on the concerted coordination or synchrony of smaller groups of neurons, whose electrical effects can be reflected in response to a specific stimulus, producing an *event related potential* (ERP). These ERPs have shown reliable individual differences that are correlated with individual differences in psychometric *g*. Thus communication between different regions of the brain required for any coherent perception, cognition, or memory consists essentially of synchronized neural activity and periodicity, or oscillations, of action potentials in the neural circuits of the brain. The most commonly identified waves found in different regions of the brain and in various states of sleep or consciousness also differ in their mean periodicity, or rate of oscillation, measured in cycles per second (cps): delta waves (<3 cps), theta waves (4−7 cps), beta rhythm (13−35 cps), gamma rhythm (30−100 cps), and alpha frequency (8−12 cps). The alpha frequency increases with age and there are large individual differences in alpha waves. Also, there is evidence for a neural pacemaker that limits the detection of temporally separated events, with a "neural window" of about 30−40 ms (Poppel, 1994). The main point here, in brief, is that the concept of periodicity or oscillation of neural impulses is not at all unrealistic or far-fetched.

What is assumed to oscillate in synchrony among a large number of neural impulses is their level of *excitatory potential*, or the momentary probability that an external stimulus will set off a train of neural impulses that is synaptically transmitted between neurons. A stimulus or a particular synaptic connection is maximally effective at the peak of the wave of excitatory potential and least effective at the trough of the wave, which can be imaged as a sine curve. The trough includes the neural refractory period in which excitatory potential is zero.

The theory posits that intraindividual variability in RT (i.e. RTSD) reflects the length of the period (i.e., the rate of neural oscillation of the sign wave of an excitatory potential. A longer period (i.e., slower oscillation) results in a larger RTSD.

The theory also posits consistent individual differences in the rate of oscillation across reliably measured RTs in different ECTs. In other words, oscillation rate is a stable characteristic of individual differences that produces positive correlations among all reliable measures of RTSD (or RTm) obtained from various ECTs. The size of the correlation between RTs on any two ECTs measured with an equal reliability depends on the tasks' complexity, because more complex tasks involve more neural "choice points" at which the effect of individual differences in oscillation rate are manifested. This increases the reliability of the RTSD, much as increasing a population sample increases the reliability of any of the sample statistics.

Individual differences in RT as a function of differences in task complexity reflect the multiplicative effect of stable individual differences in *oscillation rate × task complexity* (i.e., number of neural choice points required by the task). The empirical consequence of this multiplicative effect of *oscillation rate × task complexity* is most strikingly displayed in Brinley plots (shown in Chapters 5 and 6).

The oscillation rate affects not only the time that the initial excitatory potential is activated by the RS, but also all the subsequent interneural communication that intervenes between the occurrence of the RS and the subject's overt response, which comprises the measured RT on a given trial. The synaptic connections within the many brain pathways involved in performing a given ECT are also subject to the same consistent individual differences in oscillation rates.

Figure 11.1 illustrates how differences in RTSD result from differences in the rate of oscillation. The four waves (A−D) have different oscillation rates, so that within a given interval of time (e.g., $t_1 - t_2$) there are more peaks above the threshold of response evocation (the horizontal lines) for the waves with faster oscillations. Over many test trials, a higher rate of oscillation increases the overall probability of making faster responses to the reaction stimuli (i.e., $RS_1 - RS_4$) presented at arbitrary intervals over a given number of trials. A faster rate of oscillations necessarily results in lesser variance (hence lesser RTSD) between RTs on successive trials than do slower oscillations.

Also, when a stimulus occurs during the below-threshold phase of the wave, the excitatory phase reoccurs sooner for the faster waves, while slower oscillation results in a greater frequency of relatively slow RTs, which is manifested as greater Sk of the total distribution of RTs. Further, when the RS happens to occur during the excitatory phase in either a slow or a fast wave, it makes no difference in the RT. (In Figure 11.1, the $RS_2$ occurring at $t_2$ would produce nearly the same RT for individual waves A, C, and D, but it would produce a relatively longer RT for wave B.) Thus, the shortest RTs reflect trials on which the occurrence of the RS coincides with the wave's excitatory phase and are virtually the same for both fast and slow oscillation rates. The longest RTs reflect trials on

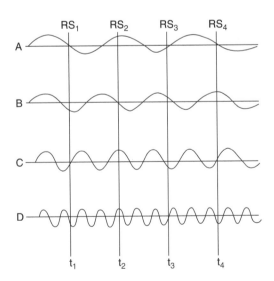

Figure 11.1: Waves of neural excitatory potential with different oscillation rates for "individuals" A, B, C, and D. RS occurs at four arbitrary points in time (*t*). (See text for a detailed explanation.)

which the RS occurs during the refractory phase of the oscillation, which also produces a greater inter-individual variance of the longer RTs. This result is reflected in the "worst performance rule" — the longer RTs produced by individuals in a given number of test trials are the most highly correlated with measures of $g$. Hence stable individual differences in the rate of neural oscillation is the basic phenomenon involved in individual differences in RTSD, general speed of information processing, and psychometric $g$. A further, even more fundamental, aim of the reductionist program of research on intelligence is to identify at the neurophysiological level the basic physical cause of stable individual differences in the hypothesized oscillation rate.

All steps in processing information in a given task take proportionally the same amount of time for a given individual. Over many tasks the average benefit of a faster rate of oscillation (hence faster speed of processing) becomes conspicuous. Therefore, as previously mentioned, individuals (and groups) show a linear relationship of RT to tasks varying in complexity, as is clearly evidenced in Brinley plots.

### Advantages of Faster Oscillation Rate in Higher Cognitive Processes

How does it come about that individual differences in oscillation rate and RT are causally connected with individual differences in conventional untimed psychometric test performance and the $g$ factor? Some of the items on such tests take not just seconds to solve, but often many minutes, and some problems are failed even when no time limit is imposed.

The explanation takes account of several principles in cognitive psychology:

(1) The focus of attention is a one-channel information processing system with a limited capacity, which restricts the amount of information that can be processed without storage and retrieval from short-term memory (STM).

(2) The information to be processed may come from external stimuli or from internal sources (i.e., STM or long-term memory (LTM)).

(3) Rapid decay of stimulus information severely limits the time that the incoming information is available for operations to be performed on it.

(4) The loss of information essential for a correct response results in a breakdown of information processing, unless the needed information can be temporarily held in STM.

(5) The process of storing or retrieving information in memory itself takes some time, which is therefore usurped from the time available for processing the incoming information. Overloading the system results in a breakdown, that is, a failure to grasp all the essential aspects of the problem needed for its solution.

(6) Therefore, individual differences in the speed and consistency of processing incoming information determines the individual's point of breakdown on the continuum of task complexity. With a slower processing rate, the task information has to be re-entered into the processing system and more of its elements have to be temporarily held in working memory (WM) and then retrieved to make a valid response (i.e., one based on all the essential information required by the task). A faster oscillation rate allows more of the elements of task relevant information to be processed without straining the capacity of WM to the point of breakdown, at which response errors occur. It is

those RTs that occur just before the point of breakdown that have the largest correlations with *g*. This is because these prebreakdown RTs reflect both the rate of oscillation and the storage capacity of WM.

### The Relationship between Working Memory and RT

The term *neural plasticity* subsumes all forms of memory, or the retention of experience. It is the fact that the nervous system retains some aspects of sensory stimulation for some length of time beyond the actual duration of the physical stimulus. The concept of *neural plasticity* in relation to *g* theory has been most fully expounded by Dennis Garlick (2002).

Besides neural oscillation, neural plasticity is the other basic process in the present theory. The relative importance of oscillation and plasticity as interacting sources of individual differences in *g* is not yet established, despite the strong claims favoring WM that are made by some cognitive theorists (e.g., Conway, Kane, & Engle, 2000). In recent years, the relationship of WM to psychometric *g* has come under extensive review and debate (Barrett, Tugade, & Engle, 2004; Ackerman, Beier, & Boyle, 2005; Beier & Ackerman, 2005; Kane, Hambrick, & Conway, 2005; Oberauer, Schulzem, Wilhelm, & Süß, 2005). Although this literature is recommended as highly informative, it does not tackle the questions raised here concerning the causal primacy of the oscillation construct in accounting for *g*. Probably, the main difficulty in achieving definitive empirical evidence on this point is that measures (such as RTm and RTSD), which reflect individual differences in oscillation rate, on the one hand, and measures of various types of memory, on the other, are so very intimately, if not inextricably, related to each other. Although in theory, individual differences in speed of processing (oscillation rate) and in neural plasticity are distinct concepts, in reality, the variance in all kinds of cognitive performance reflects, in varying proportions, the conjoint effects of oscillation and plasticity.

Plasticity is empirically distinguishable as three main memory systems that differ in length of retention and amount of information capacity: *sensory* memory (250 ms), *short-term* memory (STM<20 sec), and *long-term* memory (years). The capacities of sensory memory and LTM are indefinitely large, that of STM is comparatively very small (about four or five items or "chunks").

WM, as the active function of STM, is also of similarly limited duration and capacity. A theoretical construct introduced by Baddeley (1986), WM is a system that temporarily holds some information in STM while simultaneously processing other information coming from the sensorium or retrieved from LTM. The usual chronometric paradigm reflecting WM is a dual task. This requires the subject to perceive and retain the first-presented reaction stimulus (RS A) without responding to it, then respond to the second-presented item (RS B), and then finally respond to the first RS (RS A$^*$). For example, using a binary response console (with keys labeled YES or NO), a dual task would go sequentially as follows (only the boldface numbers appear on the display screen):

Time 1: Preparatory signal ('beep' of 500 msec)
Time 2: Blank display screen (1 sec)
Time 3: RS A: (750 msec duration)

**3+4 = ?**
Time 4: RS B: (terminated by subject's response)
**5+7−6 = 6**
Subject responds YES, with RT of 700 msec.
Time 5: RS: A$^*$ (terminated by subject's response)
**? = 5**
Subject responds NO, with RT of 800 msec.

Typically, the RTs for both tasks (Times 4 and 5) are longer in the dual task condition than when each task is presented singly. The interposed task (at Time 4) makes a demand on WM, which is reflected in the difference in the mean RTs for each of the tasks when they are presented singly and when presented in the dual task condition. Importantly, the RTs in the dual task condition are more highly g-loaded. Increasing task complexity increasingly engages WM, thereby lengthening RT until the point of a breakdown in information processing, which results in a response error. The way an information overload breakdown can be overcome without resorting to paper and pencil or a computer, is by *rehearsal* of the essential input, which consists of self-controlled repetition or some form of transformation of the information, such as chunking or mnemonic encoding (e.g., expressing 1024 as $2^{10}$ or 720 as 6!). Rehearsal is the principal means for getting information into LTM. But in our theory, the rate of rehearsal, or spontaneous recycling of sensory traces, involved in getting information into LTM is itself a function of the same rate of neural oscillation involved in RT. In effect, oscillation speed amplifies the capacity of WM by a multiplicative factor in which there are consistent individual differences. The first-order factors derived from RT tasks and WM tasks are correlated between .40 and .60 in various different data sets (Colom, Rebollo, Palacios, Juan-Espinosa, & Kyllonen, 2004).

Psychologists Lehrl and Fischer (1988, 1990), in the Erlangen University in Germany, have formulated the *capacity* of WM (*C*) in terms of a multiplicative function of the *speed of processing* (*S*) and the *duration* (*D*) of accessible stimulus traces in STM:

$$C \text{ bits} = S \text{ bits/sec} \times D \text{ sec.}$$

Individual differences in the empirically estimated value of *C* is a stronger predictor of psychometric g than is either *S* or *D* alone, but *C* and *S* differ only slightly (Kline, Draycott, & McAndrew, 1994). The g loadings of these parameters were *C* = .92, *S* = .84, and *D* = .49. Further evidence of the close association between STM capacity and RT is explained in Chapter 12.

Because the theory posits that oscillation processes both of RT and of WM in rehearsing and storing information in LTM have a common oscillation rate, measures of RT and WM should be substantially correlated. They are not perfectly correlated, however, because actual measures of RT and WM each have their own first-order factor and their specificity in addition to their common factor, which shows up in their substantial loadings on g. Measures of both STM and LTM are known to have a very considerable domain and content specificity. But individuals gain information at different rates in whatever domains they are exposed to. Hence there is a high g loading of psychometric tests of "general information." Also, there is a substantial correlation between a general RT factor and a comprehensive test of scholastic achievement, for which the individuals have had much the same opportunity for acquisition. A striking example of this phenomenon is described in Chapter 9 and Figure 9.2, which compares the RTs of groups of average and intellectually "gifted" youngsters with the RTs of university students on eight different ECTs. The RTs of the gifted youngsters and the regular university students do not differ significantly on any task. The average youngsters, however, differed significantly from the other two groups on every ECT. Most noteworthy is that the gifted and university groups, which differed 7 years in age, obtained virtually the same raw scores on the Scholastic Assessment Test and the Advanced Raven Matrices. Also, the gifted students got high grades in their university courses. By age 13, they evidently had acquired about the same amount of general knowledge and the same level of cognitive skills as possessed by the regular undergraduates, who were 7 years older, in a highly selective university that admits only the top 12.5 percent of high school graduates, based on GPA and SAT scores. Hence, it appears that individual differences in the rate of information processing, acting over a considerable span of time, can result in marked individual differences in the amount and complexity of the cognitive skills and knowledge acquired in the course of living.

### Convertibility of RT into Error Rates

Are individual differences in untimed psychometric test scores of the same casual nature as individual differences in RT tasks? The theory posits that they are essentially the same. Response errors result from information overload that strains the capacity of WM to the point of breakdown, and individual differences in this point for a given task are determined jointly by individual differences in the speed of information processing and the capacity of WM.

A paradigm for empirically demonstrating this phenomenon based on a semantic verification test (SVT, described in Chapter 2, p. 20), was presented in Chapter 9 (pp. 167–168). It showed that the difficulty levels of 14 different SVT items as measured by their mean RTs in university students are highly correlated (Pearson's $r = .70$, Spearman's $\rho = .79$) with the same 14 SVT item's error rates when the SVT was administered as an untimed paper-and-pencil test to elementary school children aged 8–9 years. The SVT items with longer RTs for the university students had larger error rates for the children, as shown in Figure 11.2. This demonstrates the convertibility of RT and untimed error rates as indices of item difficulty or complexity. Response errors indicate a breakdown in processing, and some of the SVT items are evidently beyond the processing capacity of many third graders (ages 8–9 years) even when no time limit is imposed. It is likely that the error rates for adults taking an untimed test like the Advanced Raven Matrices are of the same nature; some items exceed the processing capacity of bright university students, and in fact the mean times taken for correctly answered Raven items under a nonspeeded condition in a group of college students is directly related to the mean item error rates based on a group of university students.

Error rates on items in many untimed but highly *g*-loaded tests, such as vocabulary and general information in the Wechsler IQ tests, have other determinants than just speed of processing. The meaning of a given word or the item of information called for may never have been encountered or acquired in the first place. Or if it had been acquired it may have been forgotten and is not accessible to immediate memory at the required time. Yet the scores on such untimed *g*-loaded tests calling for past-learned knowledge are correlated

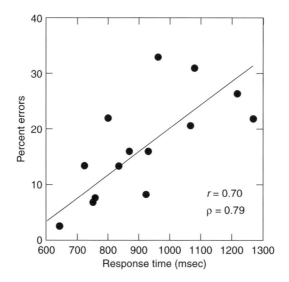

Figure 11.2: Percent errors on 14 items of the SVT in school children (aged 8–9 years) plotted as a function of the RTs to each item in university undergraduates. Data from Paul (1984).

with various measures of RT. The reason, first of all, is that individuals who process information more quickly consolidate and store more of the information to which they are exposed in LTM per unit of time than do individuals who process the available information at a slower rate, and so less of the information input is consolidated for long-term retention. Further, the acquisition of many types of knowledge content depends on inference, or relating a new item of information to old information accessible in LTM. Most of an individual's vocabulary, for example, is not acquired by deliberately memorizing the definitions of words, but by inferring the meaning of newly encountered words from the contexts in which they occur. The speed with which this kind of information processing occurs largely determines the extent of an individual's vocabulary.

Vocabulary tests, therefore, are among the most highly $g$-loaded tests. They are highly correlated with many wholly nonverbal tests that also involve some form of inference. In general, the RTs in ECTs that call for virtually no knowledge content are most highly correlated with the more $g$-loaded psychometric tests, regardless of their particular knowledge content. And when $g$ is partialled out of the substantial correlation $(-.50$ to $-.70)$ between psychometric test scores and RT, the $g$-partialled correlation drops to near zero.

Still another avenue through which processing speed comes into play is in the speed of retrieval of information from LTM, such as measured by RTs in the Posner paradigm and its variations (see Chapter 2, pp. 20–21).

Psychometric test items that call for specific knowledge content acquired by the subject in the past and must be accessed in LTM are said to measure *crystallized* intelligence, or *Gc*, in contrast to items that depend little if at all on accessing any specific content in LTM but requires on-the-spot inference or problem solving, are said to measure *fluid* intelligence, or *Gf*. The vocabulary and general information subtests of the Wechsler scales are typical examples of *Gc*, and Block Designs and Raven Matrices are examples of *Gf*. In large subject samples comprising individuals with a common language and similar schooling, *Gc* and *Gf*, are very highly correlated, and when they are residualized in a hierarchical factor analysis of a wide variety of tests, most of the variance in *Gc* and virtually all of the variance in *Gf* is "absorbed" by the single third-order factor that tops the factor hierarchy, namely, Spearman's *g*. The correlations between untimed psychometric tests and the RTs in ECTs result from the fact that they both reflect one and the same *g* factor.

# Note

1. The coefficient of reliability of the difference ($r_d$) between variables $x$ and $y$ is: $r_d = (r_{xx} - r_{yy} - 2r_{xy})/(2 - 2r_{xy})$, where $r_{xx}$ and $r_{yy}$ are the reliability coefficients of $x$ and $y$, and $r_{xy}$ is the correlation between $x$ and $y$. Measures of Sk are related to either a *difference* (between the mean and the median of the distribution) or to a *ratio* (between the third and second moments of the distribution, i.e., $\Sigma (x - x)^3 / \Sigma (y - y)^2$). Variables consisting of either differences or ratios are similarly afflicted by their liability to low reliability. For example, consider the above formula for $r_d$, assuming $x$ is the mean and $y$ is the median of an individual's RT distribution over $n$ trials. Typical figures for the reliability of the mean and the median of RT are about .90, and typical values of the correlation between the mean and the median are also about .90. When these figures are used in the formula for $r_d$, the result yields a near zero reliability coefficient for a measure of Sk, whether it is based on a difference or a ratio. As the correlation between the component elements approaches their reliability, the reliability of their difference ($r_d$) approaches zero. If our theory posits that error-free measures of the mean, median, variance, and Sk of an individual's RT distribution all reflect a single unitary phenomenon, it follows that the index of Sk would have zero reliability. To make matters worse, the variables ($x$ and $y$ in the formula) that enter into the formula for the reliability of the Sk index have correlated error components, because both measures are derived from the same set of data. These reliability problems can be largely obviated, however, by testing hypotheses regarding the relation between Sk and some external variable simply by aggregating subjects so as to create subgroups that differ clearly on the external variable (e.g., IQ). The hypothesis, then, is tested by comparing the index of Sk based on each group (rather than on individuals) with the groups' means on the external variable. This, in effect, exchanges the measurement error of individual data (which is averaged out in the group means) for the sampling error of the means, which is entirely tractable as a function of sample size. If our theory is correct in positing that Sk reflects no source of variance in RT that is statistically independent of RTm, RTmd, or RTSD, then of course there is no practical need for measuring the Sk of RT in individuals.

Chapter 12

# The Relation of RT to Other Psychological Variables

Various reaction time (RT) paradigms have also been used as a tool in the experimental analysis of psychological variables besides individual differences in the normal range of intelligence. Their focus is usually on the study of more basic psychological processes *per se*. Historically, the analysis of individual differences in these processes has less often occupied center stage. The most comprehensive literature reviews focused mainly in the traditional domain of experimental psychology are provided by Posner (1978) and Welford (1980a,b). But all these uses of chronometric methods, whether experimental or differential, are made possible because the mental processes of interest are intrinsically time dependent at the level of basic causal mechanisms. This is easily overlooked when, at a superficial level of observation, the subject's performance does not appear to be speeded or time dependent. Typically, however, processing time is itself part and parcel of the specific cognitive process under chronometric analysis.

## Short-Term Memory Span

Memory processes are a good example of this. They have been subjected to chronometric study, probably more than any other class of variables (McNicol & Stewart, 1980). Various studies include short-term memory (STM), long-term memory (LTM), and retention loss following delayed recall and the effects of proactive and retroactive interference.

Scientific advances consist, in part, of discovering constants, ratio properties, or invariance in the quantitative features of a given phenomenon. This is illustrated in the recently discovered properties of the *memory span*. Memory span is one of the earliest and simplest tests of STM capacity (Dempster, 1981). It was included in the first intelligence test by Binet in 1905 and has been retained in present-day IQ tests. Though it is not typically administered as a timed test in psychometric batteries, its chronometric features, which have come under recent study, show that memory span is not in the least a trivial or superficial variable in the realm of cognition. Its comparatively poor reputation as a subtest in IQ batteries, such as the Stanford–Binet and the Wechsler tests, results from psychometric and statistical artifacts. These result in the memory span subtest's having a considerably lower *g* loading than most of the other subtests included in such batteries. The reasons are several:

(1) An individual's memory span is generally reported on a *cardinal scale* (e.g., the largest whole number of presented items that can be correctly recalled on 50 percent of the trials), which is a much too crude scale for accurate estimation, dividing virtually the total range of memory span in the population into a discrete 10-point scale, with the vast majority falling around the famous "magical number $7 \pm 2$" (Miller,

1956). An individual's memory span oscillates from trial to trial, so the mean (or median) of the distribution of the individual's spans over *n* trials provides the more precise measure.

(2) The *low reliability* because of too few trials in the span test.

(3) The high-content *specificity* of the span test when it is based on only one type of stimuli (e.g., digits).

(4) *Oscillation* of response strength at the threshold of recall, which is more prominent in the more elementary cognitive processes, thus affecting the reliability of STM more than of LTM as represented in the most highly *g* loaded subtests, such as vocabulary and general information.

Therefore, the construction of a battery of memory span subtests consisting of diverse stimuli (e.g., digits, letters, words, shapes, colors), with at least 10–20 test trials on each type of stimuli would markedly increase the reliability, diminish the specificity, and increase the *g* loading of the composite memory span score. These conditions are critical for the study of individual differences in STM variables, such as memory span because the *time dependency* of memory span is the basis of its correlation with psychometric *g*.

Before presenting the evidence for this hypothesis, however, the essential constancies of individual differences in memory span should be noted.

(1) Memory span is a unitary variable across visual and auditory modalities. There is a perfect disattentuated correlation between *visual* and *auditory* memory span for digits (Jensen, 1971).

(2) There is a perfect disattenuated correlation between (a) *memory* scanning rate for digits held in STM (i.e., the Sternberg paradigm) and (b) the rate for *visually* scanning the same number of digits displayed visually (i.e., the Neisser paradigm, described in Chapter 2, p. 22) (Jensen, 1987a,b). This implies that the same time-dependent processes (i.e., scanning rates) govern both the rate of initial visual intake of information into an individual's STM and the rate of its recall from STM.

(2) The rate that individuals can read aloud a presented list of numbers or simple words predicts their memory span for the same type of stimuli. It turns out that individuals can recall as many random words in STM as they can articulate on average in about 1.8 s, regardless of the number of syllables in each item (Baddeley, Thomson, & Buchanan, 1975). This implies a time constraint on the rate at which the intake can be rehearsed before time-lapse decay of the memory trace or retroactive interference from any further input that exceeds STM span.

Repetition or rehearsal of the information input through what Baddeley has termed the "phonological loop" consolidates fresh information for later recall, which is related to the rate of speed with which individuals can verbally articulate the items to be later recalled. This rehearsal speed of the stimulus input, therefore, is a time-constrained function of working memory by which the traces of newly input information in STM is consolidated. This is well illustrated in a large-sample developmental study by Kail and Park (1994) of children 7–14 years of age in the United States and Korea. A causal modeling analysis was based on the composite scores of two different tests for each of the three experimentally independent variables: (1) speed of information processing, and (2) speech articulation rate (for digits and letters). The best-fitting causal model indicates that, for both the United States and Korean samples, the age-related increase in memory span (for digits and letters) is mediated by age-related changes in information processing speed, which in turn governs vocal articulation speed. Because the speed-of-processing tasks themselves make minimal demands on either verbal articulation or STM, processing speed has primacy in the causal pathway (i.e., processing speed ➔ articulation speed ➔ memory span).

The same overall conclusion was drawn in more general terms in the most comprehensive, thoroughly analytical, and integrative review of the evidence on the relationship between processing speed and working memory in children that I have found — a "must read" prize-winning article. The authors stated that "…most of the effect of the age-related increase in [processing] speed on intelligence appears to be mediated through the effect of speed on working memory. Finally, most of the effect of the age-related improvement in working memory on intelligence is itself attributable to the effect of the increase in speed on working memory, providing evidence of a cognitive developmental cascade" (Fry & Hale, 2000, p. 1).

The very lawful generality of memory span data across highly diverse item contents is seldom appreciated. A classic meta-analysis of span data based on seven distinctly different classes of stimuli used in 45 independent studies shows that a constant information processing rate is the basis for the observed regularity or lawfulness of the memory span data (Cavanagh, 1972). The analysis makes use of the Sternberg memory scan (MS) paradigm (see Chapter 2, pp. 21–22), in which a subject is presented a short list of items of a particular type, then is presented a single probe item of the same type, to which the subject responds "yes" or "no" depending on whether the probe was or was not a member of the presented list. The subject's RT, then, reflects the processing rate required for scanning the memorized list of items for the presence or absence of the single probe. The related datum of interest is the typical measure of memory span for the particular class of stimuli namely, the largest number of stimuli that can be recalled after one presentation (at a rate of 1–2 s/item) on 50 percent of the trials. The span for any particular class of stimuli is unitized by taking its reciprocal. All the subjects in the various data sets represented in this meta-analysis were adults, and all the span stimuli were presented only visually. Figure 12.1 shows the near perfect linear relationship between the mean processing rates (millisecond per item) for the given stimuli and the reciprocals of the memory spans for the same types of stimuli. Since the data points here are group means, it would be of considerable interest to determine the degree to which individual data fit the linear function. Cavanagh notes that "the [average] time required to process a full memory load is a constant, independent of the type of material stored, with a value of the order of one-quarter of a second, the slope of the regression line [243.2 msec] in Figure 12.1" (p. 529). Conversely, the faster the processing time for any given type of stimuli, the greater is the memory span for such stimuli. The mean memory span for different classes of stimuli differs depending on their complexity (i.e., the mean number of features per item that need to be stored for correct immediate recall); the memory span therefore is smaller for more complex items. Thus, memory span and processing rate are both measures of the same processing system. The storage capacity of STM as represented by the memory span can be conceived basically as a time-dependent process governing the number of items that can be reported before the memory trace decays.

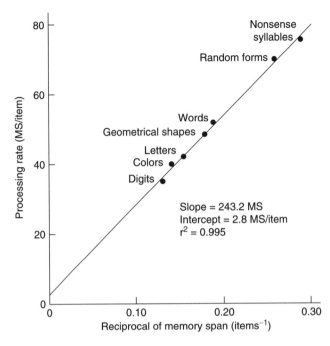

Figure 12.1: The linear regression of processing (i.e., STM scanning) rate on the reciprocal of the memory span for seven classes of stimuli. (From Cavanagh, 1972, with permission of the American Psychological Association.)

## Long-Term Memory, Consolidation, Retrieval, and Forgetting

When the amount of information to be retrieved exceeds the immediate memory span (for a given type of stimuli), rehearsal of the input over repeated trials becomes necessary. At this point, we cross the fuzzy boundary between STM and LTM. It is exemplified by the classical serial rote-learning paradigm in which the subject must memorize (up to a given criterion of accurate recall of the items in the presented order), such as a list of words, numbers, nonsense syllables, or other stimuli that clearly exceeds the subject's immediate memory span. This acquisition process is highly time dependent in two fundamental ways: (1) the item learning rate (and hence the number of trials needed to attain a given criterion of mastery) is systematically affected by the experimenter-controlled presentation rate of the items on each learning trial; and (2) in the absence of interference occurring shortly after the last learning trail, and without further practice, the newly learned material becomes increasingly consolidated in an LTM store through an autonomous physiological process that strengthens the neural memory trace over a number of hours. Thus, LTM actually refers to two distinct but closely related phenomena: (1) the initial learning and retention of material that exceeds the capacity of STM, and (2) the autonomous neural consolidation or preservation of memory traces so that they can be accessed over a very long period of time, and even long after the initial consolidation their strength can be increased by further practice or rehearsal. It has long been experimentally established in animals and humans that this autonomous time-dependent biological process of long-term neural consolidation is basic to learning and long-term retrieval (McGaugh, 1966).

A chronometric method was used to discover that the process of retrieval fundamentally differs for items within the span memory and for the same kinds of items in super-span lists that characterize serial rote learning. This was shown by the marked difference in retrieval rate of items in the two conditions when they are subjected to the Sternberg paradigm (Chapter 2, pp. 21–22). In a study by Burrows and Okada (1975), subjects memorized lists of words to a criterion of two perfect trials. The lists varied in length from 2 to 20 words. Retrieval rate is a bilinear function of list lengths distinguishing STM and LTM, being very much slower for lists within the range of memory span (i.e., 2–6 words) than for super-span lists (i.e., 8–20 words), as shown for two experiments in Figure 12.2. (The different procedural conditions affecting overall task difficulty in experiments 1 and 2 are not germane to the point being made here about the difference between STM and LTM.) A detailed critical discussion of this study, however, questions its generality, noting other studies in which retrieval (scanning) rates are not unilinear even within the range of the STM memory span (McNicol & Stewart, 1980, pp. 279–281).

Although there are probably different acquisition and retrieval processes for STM and LTM, the correlation of individual differences between them or their relationship to other

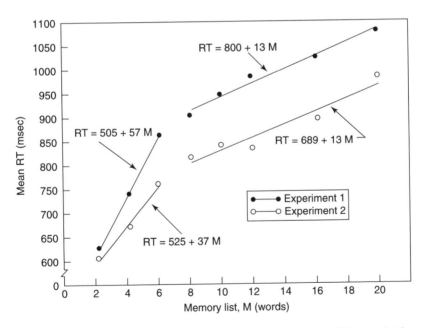

Figure 12.2: The best-fitting bilinear regression of retrieval time (milliseconds) for memorized words in short (2–6 items) and long (8–20 items) lists in two experimental conditions that affect overall difficulty level. (From Burrows & Okada, 1975, with permission of Science.)

variables, such as age, IQ, or various brain pathologies has not yet been systematically investigated under uniform methodological conditions.

The well-known Posner paradigm (Posner et al. (1969) and its variations (see Chapter 2, pp. 20–21) has been the typical chronometric method for studying the speed of retrieval of information from what is clearly LTM. The main feature of this paradigm is that it contrasts two kinds of RT: (1) the RT for discriminating between a pair of stimuli that are either *physically* identical or physically different (known as the PI condition), which makes virtually no demand on LTM, and (2) the RT for discriminating between a pair of stimuli the *names* of which are semantically either the same or different (the NI condition) which demands the retrieval of information from LTM. The difference in mean RTs (NI–PI) indicates the time required to access highly overlearned information in LTM. A classic study of individual differences using the original and simplest form of this paradigm (where the paired stimuli are single upper-case and lower-case letters) found that the mean NI–PI difference in RT discriminated between groups of college students that differed on a standard verbal test of intelligence (Hunt et al., 1975).

In a more demanding version of the Posner paradigm, the PI condition has pairs of words that the subject must distinguish as physically the same or different (e.g., dog–dog versus dog–fog), and NI condition has paired words that are semantically the similar or different meanings (e.g., evil–bad versus evil–good). In a related paradigm that depends on retrieval of information from LTM, the subject discriminates (Yes or No) between pairs of words that belong to the same category or to different categories (e.g., chair–bed or apple–lake).

The latency of retrieval varies for different items of general information held in LTM. This phenomenon can be used to analyze the organization of the LTM storage system and the strategy of accessing LTM for the recall of information. The subject simply responds True or False to a variety of ostensibly simple information items, such as "President Lincoln was shot" or "The Mississippi River is in Southern India." As the main variable of interest is the speed of accessing items most likely to have at one time entered the subject's LTM, the items are selected to ensure low error rates for the class of subjects being tested. Using this technique in experiments in which knowledge acquisition is controlled by teaching novel "facts" and measuring their retrieval latencies, John Anderson (1983) has discovered features of recall latency that he explains in terms of a spreading activation ("fanning") through the neural network to retrieve the specific information called for. It involves neural pathways fanning out from a more general conceptual node to the more specific information being searched for. Anderson's "spreading activation" model and its associated chronometric methods could be fruitfully applied to studying what we know to be the very wide range of individual differences in the retrieval of information in LTM; particularly, differences associated with cognitive aging and with various brain pathologies. Because the strength of memory traces is not an all-or-none phenomenon, RT is an especially sensitive-graded measure of memory.

The effects of time elapsed between initial reception, and the rates of increase in both proactive and retroactive inhibition of memory traces are very sensitively registered bythe *continuous recognition memory* paradigm (McNicol & Stewart, 1980, p. 283). The subject is presented a continuous list of stimuli of a given type (e.g., words, nonsense syllables, numbers, symbols, etc.) one at a time. On each presentation, the subject leaves the home button and presses one of two response buttons labeled NEW (if the item is encountered for the first time in this list), or OLD (if previously encountered in the list). Following the subject's response another item appears on the screen. The sequential intervals between new and old items are typically varied throughout the sequence. Both RTs and errors for new and old items are registered in terms of their sequential intervals, which reflect the effects of proactive and retroactive inhibition of retrieval as well as the effects of number of previous exposures and the time elapsed since the $n$th exposure. This method so far has been little used in individual differences research except for the detection of brain pathology.

Access to vocabulary, too, is not an all-or-none variable. It can be assessed as a continuous variable chronometrically by presenting a sequence of fairly common words of graded difficulty (e.g., in terms of their population $p$ values or their Thorndike–Lorge word frequencies) randomly interspersed by a given proportion ($1/2$, $1/3$, etc.) of paralogs (i.e., pronounceable letter combinations that merely look much like words yet are not found in any unabridged English dictionary or dictionaries of slang, e.g., 'byscenity'). The subject responds to buttons labeled YES or NO as to whether the given item is an authentic word. RTs and errors are recorded on each trial.

## Automatic Retrieval of Basic Skills from Long-Term Memory

The automatizing of a basic cognitive skill involves short-circuiting its defining operations, thereby greatly quickening its retrieval. The sheer recall of simple arithmetic number facts, such as adding, subtracting, or multiplying single-digit numbers, which are memorized and practiced repeatedly requires a conceptually much less complex level of cognition than understanding abstractly the mathematical operations that define addition, subtraction, and multiplication. It is important to know these operations so that they can be applied when needed in a given problem. That granted, it is also useful, and often crucial, to have speedy access to specific "number facts" stored in LTM, especially in solving complex thought problems, in which it is advantageous when several of successive operations can be performed within the time span of working memory. Bright pupils, who otherwise have no problem grasping mathematical principles, can have difficulty with arithmetic thought problems and also with acquiring the higher levels of proficiency in math. The explanation is usually that they have not sufficiently automatized the retrieval of basic tools, such as simple number facts stored in LTM (Bull & Johnston, 1997). The same principle applies to acquiring automatic recognition of mathematical symbols and commonly occurring components or symbol clusters in math and statistical formulas, phonemic decoding of printed words, and musical notation, to name a few. Elementary education in recent years has focused so strongly on conceptual learning and has so depreciated rote memory and drill that many pupils are left without the automatized skills needed for further advancement in math beyond the elementary grades. Even performing long division and fractions, typically taught in grade 4, becomes almost impossibly difficult for some pupils because of insufficient automatization of the more basic arithmetic skills.

Chronometry affords the sensitivity and precision required to assess the degree to which basic skills have been automatized. This can seldom be detected by ordinary PP tests with right/wrong item scoring. A chronometric paradigm called the math verification test (MVT), however, reveals a wide range of individual and group differences in RTs to basic number facts that usually are mastered before grade 4 (Jensen & Whang, 1994). Such RT differences were revealed by the MVT even among pupils in grades 4–6 all of whom achieved 100 percent perfect scores on a liberally timed PP test covering the very same basic skills as assessed by the MVT. There are still large individual differences in automatization of number facts, as well as consistent intraindividual variation in the accessibility of different number facts. The MVT is a RT measure of speed of accessing elementary number facts from LTM: addition, subtraction, and multiplication of single-digit numbers. For example, using a binary response console with a home button (see Figure 2.4, p. 19), the subject responds (YES or NO) to items such as $2 + 4 = 6$; $7 - 2 = 4$; $2 \times 3 = 5$, etc. Retrieval speed on the MVT is related to Raven Matrices Scores (as an index of *g*) and, independently of *g*, to performance in more advanced levels of arithmetic as measured by standardized achievement tests. However, the critical question that cannot be answered by a purely correlational study is the degree to which extensive practice on number facts speeds their retrieval and particularly whether this is related to later math achievement independently of nonnumerical measures of RT. The answer, of course, will have to be sought by a long-term experimental study — essentially the paradigm for measuring the transfer of training. Automatization of a certain class of skills (e.g., mathematical, verbal–linguistic, musical, mechanical, athletic) is invariably a major component in the acquisition of high levels of expertise and it is probably the one variable that most clearly distinguishes between experts and the run-of-the-mill in most fields of performance.

## Mental Retardation

For a long period during the mid-1900s when chronometric research was at its lowest ebb in general experimental and psychometrics, its empirical revival in the late 1960s occurred in research on mental retardation (MR) in hopes of finding a more sharply analytic means of describing the nature of cognitive deficits than is afforded by conventional psychometric tests.

The chief pioneers in this effort were Alfred Baumeister and co-workers. Their extensive empirical research on retardation is closely linked to the well-known variables of experimental chronometry. In many experiments using analysis of variance designs not only were very large differences found between retarded and normal groups in *main* effects (e.g., RT means and RTSDs), as expected, but also there were often significant *interactions* between groups and variations in experimentally manipulated variables, such as stimulus intensity, preparatory interval, response–stimulus interval, effects of rewards, stimulus and response complexity, and stimulus compounding (see the references Baumeister & Kellas, 1967, 1968a,b). Systematic research in this tradition was later pursued in the University of Adelaide, Australia by Nettelbeck and others (e.g., Nettelbeck, 1980; Nettelbeck & Brewer, 1976).

What is not clearly established by these experimental study designs is whether the reported significant interactions between groups (retarded versus normal) and procedural conditions represent truly dichotomous differences unique to the group difference but do not also occur *within* each group. The interaction effect in the groups × conditions design can suggest uniqueness of a group difference when the seeming dichotomous effect actually represents a continuous variable consistently running throughout the total range of ability encompassed by both groups.

A striking example is seen in a study of the Hick paradigm with severely retarded young adult subjects (mean IQ 38.5) (Jensen, Schafer, & Crinella, 1981). Besides their showing very slow RT and MT, there were two effects that differed markedly from all other groups we had tested at that time, including mildly retarded young adults (mean IQ 70), using the same RT–MT apparatus. The severely retarded did not display Hick's law and, unlike any other individuals or groups we had ever tested, their MT was greater than their RT. The severely retarded group thus appeared to be entirely unique with respect to the RT–MT difference, just the opposite of the general finding. This reversal would constitute a strong interaction in terms of the groups × conditions (RT–MT) design. Yet by looking at the ratio RT/MT for simple RT (i.e., 0 bits in the Hick test) in several other groups differing widely in mean IQ, the observed reversal effect in the severely retarded does not appear to be at all unique, as it falls on the same linear continuum that holds for other groups of much higher IQ, going from the mildly retarded (IQ 70) to university students (IQ 120), shown in Figure 12.3.

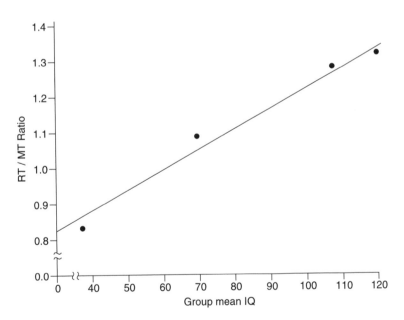

Figure 12.3: Mean RT/MT ratio for one-button condition in Hick paradigm (SRT), plotted as a function of average IQ levels of four group: severely retarded ($N = 60$), borderline retarded ($N = 46$), vocational college students ($N = 200$), university students ($N = 50$). (From Jensen et al., 1981 with permission of Elsevier.)

A remarkable analysis by Kail (1992) revealed a similar linear continuity of RTs between normal and retarded groups (mean IQs 33–72) across RT measures representing a great variety of experimental conditions that made for a considerable range of task difficulty. For each such RT test, there was a mean RT for a retarded group and a mean RT for a nonretarded group, comprising in all 518 pairs of mean RTs. The groups were stratified by age and IQ level, allowing six separate regression analyses in each of which the retarded groups' RTs were regressed on the normal groups' RTs. This analysis shows that the retarded groups' RTs across the wide variety of tasks are well predicted by a single multiple (*m*) of the normal groups' RTs, with *m* ranging from 1.4 to 2.3. The goodness of fit to the linear regression is indicated by values of $R^2$ ranging from .904 to .989 for the six groups. Only the most borderline retarded (IQs 68–71) showed a departure from linear regression, which Kail speculates could reflect differences in response strategy in this borderline group. The main conclusion of the study, however, is summarized by Kail as follows "These results are consistent with the view that differences in processing speed between persons with and without mental retardation reflect some general (i.e., nontask specific) component of cognitive processing." (p. 333).

All of the most generally agreed upon conclusions drawn from the chronometric studies of MR as clinically diagnosed by standard psychometric tests (IQ < 70) are predictable from the theory outlined in Chapter 11 (e.g., Baumeister & Kellas, 1968a,b; Liebert & Baumeister, 1973; Detterman et al., 1992; Nettelbeck, 1980; Nettelbeck & Brewer, 1976; Vernon, 1981):

(1) The psychometric differences between the retarded (R) and nonretarded (NR) are predominantly related to the tests' *g* loadings.

(2) R and NR groups always differ in *mean median* RT on every chronometric paradigm that has been used.

(3) The largest R–NR differences, measured as effect size, are found for the intrasubject trial-to-trial variability in RT.

(4) When both RT and MT are measured on various tasks that differ in complexity, R and NR differ on both RT and MT, but the RT differences are more highly sensitive to differences in task complexity and *g* loading, whereas the corresponding MT differences though considerable between R and NR groups, show hardly any sensitivity to differences in task complexity or *g* loading (e.g., Vernon, 1981). This suggests there is a slowing of both the cognitive (RT) processing and psychomotor (MT) functions in the retarded.

The main debate still going on in this field is whether the cognitive deficit associated with retardation is unitary and quantitative, as a downward extension underlying the distribution of biologically normal variation in *g*. Or is MR better described behaviorally as a number of multidimensional deficits? The unsettled issues of the debate are prominent in a major discussion on the subject set off by the report of an impressively large study by Detterman et al. (1992), accompanied by critical commentaries. As the largest chronometric study of MR ever performed, both in sample size (*N* > 4000) and the number of chronometric measures (nine distinct tasks yielding 31 variables). The authors argue that retardation, and normal variation in *g* as well, depend on what they call a *modal model*. It consists of a small number of very basic processing modules, including very short-term memory (VSTM), STM and LTM, and an Executive that governs the deployment of the lower-order modules for the execution of a given cognitive task. The modules are related temporally in terms of the order in which they come into play in processing stimulus input. Thus, if module number 1 in the temporal order is defective or inefficient, all of the higher-order modules (2, 3, etc.) are thereby handicapped. Hence, the degree of severity of the effect of a weakness in a given module depends mainly on whether the module comes into play earlier or later in the temporal order of information processing.

The large-scale study intended to test this model yielded somewhat supportive but less than compelling evidence for it. The essential problem with this study, as with so many others in this field, is the now very well-established fact that, in terms of biological etiology, there are many different causes of MR. Hence, behavioral tests of a modular theory are rendered indecisive by the blurred biological discontinuity between what is considered "normal" variation in g or IQ and the cognitive deficiencies associated with clear organic defects, whether endogenous or exogenous.

MR is typically defined psychometrically as IQ < 70. But that cut-score arbitrarily divides what is actually a mixture of various etiologies that have overlapping distributions of ability detectable at the behavioral level. Brain defects of biological origin most probably occur throughout the whole range of the IQ, but their relative frequency markedly differs as a continuous variable in different segments of the IQ scale. It increases proportionally at a greatly accelerated rate as we go toward the lower tail of the IQ distribution. (A specific identifiable biological defect [chromosomal, genetic, brain damage, etc.] is increasingly probable the lower the IQ; what is uncertain is the lowest point on the IQ scale below which the continuum of biologically normal variation ceases to be causally decremental. In other words, what would be the lowest possible IQ in the absence of biological defect?) Therefore, the average of psychometric or chronometric characteristics for a group of persons diagnosed as mentally retarded by a given cut-score on any behavioral test represents a mixture of etiologies and very likely a mixture of many distinct defects.

The precision of measurement made possible with chronometric techniques makes them an ideal means for creating a differential taxonomy of the particular cognitive deficits among persons psychometrically classified as MR. But analyzing the nature of cognitive defect in terms of information processes calls for a quite different approach than what is usually taken — that is, contrasting the means of an unselected MR group (e.g., IQ < 70) and a "normal" group (e.g., IQ > 100) on various tests. The retarded group, then, is itself too heterogeneous and can only be described as retarded in general. The chronometric search should begin with contrasting carefully selected sex- and age-matched groups with specific known etiologies. (Many types of retardation can now be identified by genetic analysis and MRI or PET brain scans.) It is important first to establish the degree of individual variability that exists even *within* the most homogenous etiological categories. The analysis of a chronometric battery of ECTs would not be based simply on mean differences between the selected etiological groups but by suitable multivariate methods that indicate the validity of grouping individuals in terms of information processes. The use of a factor analytic method known as Q-technique lends itself to this kind of typological analysis. (Q-technique is explicated in most textbooks on factor analysis.)

Q-technique is the factor analysis of a number of *individuals* each measured on a number of tests, in contrast to the usual R-technique, which is the factor analysis of a number of *tests* given to a group of individuals. As R-technique sorts various *tests* into groups (i.e., factors) in terms of the tests' degree of similarity, so Q-technique sorts *individuals* into groups in terms of their degree of similarity across a number of different tests. By means of Q-technique applied to a relatively heterogeneous group those individuals whose profile of scores on a variety of different, narrowly focused tests are most similar are sorted into much more homogeneous groups in terms of their ability *patterns* (not their overall mean scores). Hence, different subgroups of MR can be distinguished at the overt ability level even if these subgroups all had the very same mean IQ. We know there is a true etiological or physical typology for many forms of MR that involve more than just the lower extension of psychometric *g*. If there is also a clear typology of information processing abilities among persons with MR, it would be revealed by Q-technique. The degree of congruence between the processing typology and the physical typology of MR would be of great interest, mainly for the light it might throw on the physical basis of information processes within the normal range of IQ. Knowledge of the specific ways in which an inscrutably complex system occasionally fails can often provide clues about how the system normally works.

## Giftedness and Special Talents

The term *gifted* as used in an educational context commonly refers to a high level of general mental ability, or psychometric g, which is typically associated with accelerated progress in scholastic achievements. In a broader context, gifted may refer to exceptional special talents or any abilities or personality traits — artistic, musical, athletic, interpersonal — that are out of the ordinary (Jensen, 1996, 2004a; Simonton, 2003).

The criterion for academic giftedness in most school systems is a score on highly *g* loaded tests that is two or more standard deviations above the mean of the general population (IQ > 130). As such, we should expect chronometric tests to show differences between gifted (G) and nongifted (NG) pupils to the extent that such chronometric tests are correlated with psychometric *g*. What comes as a surprise to many parents and educators is that groups selected as G differ quite markedly from the much larger NG segment of the school population in their speed of information processing even when the processing tasks are very simple and have no content of a scholastic nature. The NG–G difference is clearly shown in a study (Kranzler et al., 1994) that compared groups of G and NG students (aged 11–14) on three processing tasks: simple RT (SRT), 8-choice RT (CRT), and discrimination RT (DRT) based on the odd-man-out paradigm (described in Chapter 2, p. 30). Figure 12.4 shows the results for mean RT (panel A) and for intraindividual variability (RTSD) (panel B). The NG–G differences in both the mean RT and the SDRT increase as a function of task complexity. Although the differences may appear small in absolute magnitude, in terms of effect size (ES = mean differences between the groups divided by the average SD within groups) the ES is approximately .25 on SRT, .75 on CRT, and 1.30 on DRT for both the mean RT and the mean SDRT. The long-term effects of this magnitude of differences in speed of information processing over an extended exposure to

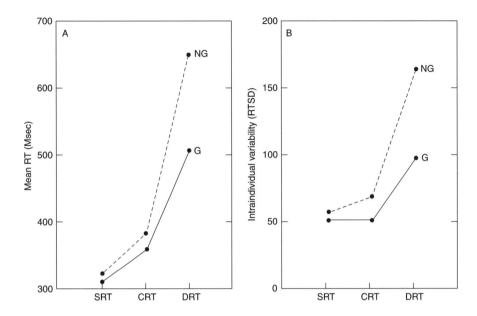

Figure 12.4: (Panel A) mean RT and (Panel B) mean RTSD of Gifted (G) and Non-gifted (NG) groups on each of three cognitive tasks: Simple reaction time (SRT), Choice RT (CRT), and Discrimination RT (DRT). (From Kranzler et al., 1994.)

many learning experiences easily accounts for the marked superiority in general knowledge and other cognitive skills and achievements accrued by G as compared with NG pupils with equal opportunities over any given period of time.

What has not yet been determined is whether the highest levels of the *g* dimension that can be convincingly assessed by standardized psychometric scales are measurable by speed of information processing on relatively simple chronometric tasks. Within the range of standardized psychometric IQ up to about 160, there seems to be no departure from a linear relationship between IQ and either CRT or DRT. More extraordinary levels of a specific kind of cognitive performance probably represent some quite specialized and highly focused investment of *g* in a very narrow sphere of cognition. Shakuntala Devi, one of the most astounding numerical calculators in the world, for example, tested only within the average range for college undergraduates on both psychometric and chronometric tests, except for the one chronometric test based on a numerical version of the Sternberg paradigm, which measures the speed of STM scanning of series of 1–7 digits. Devi's performance did not show the typical Sternberg effect (a linear increase in RT as a function of the number of digits presented), a phenomenon that has always appeared in the hundreds of undergraduates tested under exactly the same conditions (Jensen, 1990). Scanning even as few as seven digits in memory is evidently performed by some quite idiosyncratic process by a numerical savant like Devi, who could mentally calculate the 20th root of a 200 digit number in less than one minute.

### Surpassing Talent

I have not found an application of chronometric methods to the study of individual differences in special talents, such as aptitude for musical and artistic performance *per se*. The nearest thing to it is the Seashore Musical Aptitude Test and its similar offshoots, which objectively measure variables involving time, such as thresholds of pitch discrimination (measured in hertz) and durations of tones (in millisecond). The sum of such elemental aptitudes, however, do not add up to musical talent by any means, although it is highly likely that well above-average performance on such tests of auditory discrimination is a necessary-but-not-sufficient condition for a successful career as a professional musician.

The most interesting use of chronometry in this realm that I have come across, however, is in the purely objective analysis of musical performance *per se*. Music critics' evaluations of concert performances and their characterization of different musical interpretations of masterpieces in the standard repertoire have long been a literary art form. It typically describes and compares, in largely musical terminology along with evocative similes and metaphors, the critic's subjective impressions and evaluations of noted concert artists' performance of famous masterpieces. Although these highly subjective impressions have some broad consistency and validity, they also show many differences among critics. They are too broad-brushed for precise comparisons, leaving much of whatever properties differentiate musicians in a literary limbo of mystery. One of the typical questions in this realm, for example, is what distinguishes the performances of the acknowledged great and famous concert artists from those of outstanding conservatory graduates and successful professionals. One thing is clear: it is not technical skill and accuracy *per se*. Concert audiences do not fill Carnegie Hall to hear letter-perfect renditions of their favorite masterpieces. That is mere baseline expectation, and even a few technical imperfections in performance are overlooked, provided the audience experiences certain appealing qualities. The audience seeks a "musical experience," a kind of aesthetic excitement. But this is strictly auditory and it comes through on recordings as much as in an observed live performance. It has nothing to do with the performer's appearance, on-stage personality, showmanship, or publicity hype.

What are the essential features of performance itself that distinguishes the super-talents — a Kreisler and a Heifetz, a Paderewski and a Horowitz, a Casals and a Rostropovich, a Callas and a Flagstad, a Toscanini and a Furtwängler — from the performances of those excellent professionals who never attain such exceptional acclaim?

The answer seems to lie in the entirely unique and idiosyncratic distortions of the musical score in subtle ways that evoke an emotional response in most musically sensitive listeners, a kind of aesthetic excitement. A strictly accurate, literal rendition of the work hardly evokes such a response. The departures from literalness are not a flouting of the composers intentions as shown in the printed musical score, but are an idiosyncratic attempt to express the composer's inspiration. In this effort, the great performers differ much more among themselves than do the adequate but less gifted performers. An expert can immediately recognize whether the interpretation of a famous work is performed by, say, Kreisler or Heifetz, Rubinstein or Horowitz, Toscanini or Stokowski. Each of these great artists has a unique and recognizable "musical personality" that infuses virtually all

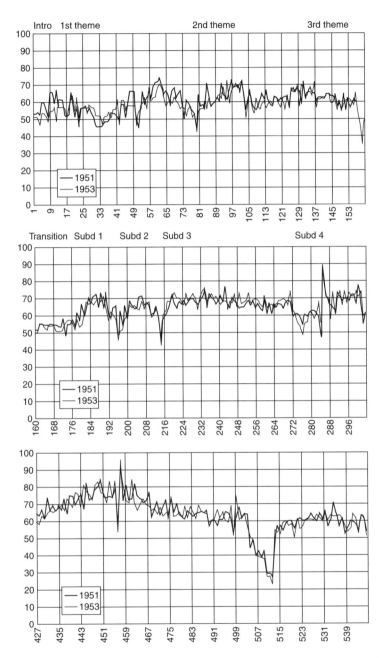

Figure 12.5: Chronometric measurement of tempo fluctuations in two performances (recorded live in 1951 and 1953) showing Furtwängler's rendition of the first movement of Beethoven's Ninth Symphony (From Rink, 1995, with permission of Cambridge University Press.)

the works in their repertoire. One of the musical dimensions in which "musical personality" is most prominently expressed is in tempo and its fluctuations.

Recently, some musicologists have subjected these purely tempo aspects of musical interpretation to entirely objective and fine-grained analysis by means of chronometric technology applied to recorded instrumental and orchestral performances by famous instrumentalists and conductors (Rink, 1995). A striking example of this is the analysis of tempo fluctuations throughout the first movement of Beethoven's Ninth Symphony as conducted by Wilhelm Furtwängler, generally acknowledged as one of the greatest conductors of the twentieth century. He was highly acclaimed for his interpretations of Beethoven's orchestral works, especially the great Ninth ("Choral") Symphony, which he recorded with several of the world's great orchestras. Although the last and most technically advanced of these recordings were made in the early 1950s, they are still in demand by the musical cognoscenti.

A brief example of the chronometric analysis of Furtwängler's rendition of the first movement of Beethoven's Ninth is shown in Figure 12.5. The ordinate indicates the tempo within each measure of the score in standard metronome units (i.e., number of beats per minute); the abscissa marks the measures of the musical score, which is in 2/4 time throughout the first movement. (Shown here are only 429 measures of the total 547 in the first movement.) Beethoven's own metronome marking printed at the beginning of the score is 88 and if performed consistently at that tempo throughout would create a graph with a straight line through all the measures at the 88 point on the ordinate, except for three places in the score where Beethoven indicates a retard in just the second half of each of these measures, which occur at different points in the first movement. Incidentally, no major conductors observe Beethoven's metronome marking of 88, considering it a mistake, possibly resulting from Beethoven's total deafness at that time. The overall average tempo of most conductors in this movement is about 75. But the most conspicuous feature in Figure 12.5 is the great variability in tempo, varying almost from measure to measure. Evidence that the tempo fluctuations are not merely haphazard or due to any lack of control of the orchestra is shown by superimposing the graphs made from two different performances two years apart (1951 and 1953) with different orchestras (Berlin Philharmonic and Vienna Philharmonic). The degree of similarity in the tempo fluctuations of these two performances is equivalent to a reliability coefficient of about .97. It is clear that Furtwängler's unique rendition of Beethoven's Ninth (as with virtually everything else he conducted) was entirely intentional and a result of a superb, almost magical, fine-grained control of the orchestral forces. This degree of inducing such expressive subtleties throughout the entire movement, however, could never be achieved merely through Furtwängler's instruction and rehearsal, but must necessarily come about almost entirely emphatically, by every player's involuntarily reflecting the conductor's individual feeling of the musical expression. Only the greatest conductors, such as Furtwängler, Toscanini and Stokowski, can induce such a wholly unique emphatic response from a great orchestra. From a purely subjective musical standpoint the tempo fluctuations that appear so conspicuous in Figure 12.5 seem hardly noticeable in the heard performance, yet they evoke the important emotional features of the music that mysteriously create its subjectively moving and heroic character, making it sound peculiarly "Beethovean."

# Chapter 13

# Clinical and Medical Uses of Chronometry

In the quarter century since Nettelbeck's (1980) selective review of the literature on pathological conditions that affect reaction time (RT), there has been a virtual explosion of the uses of RT measurements in this sphere. A comprehensive bibliography of medical research studies since 1980 utilizing chronometric methods would overwhelm the entire bibliography of this book, and an exhaustive search of the medical literature might well turn up more references to RT than that now exist in the psychological literature *per se*. For the most part, however, the many reports of the varied clinical uses of RT tests contribute little to our understanding the basis of RT correlations with other psychometric variables, except to underline the fact that RT generally is highly sensitive to many physiological conditions that affect brain functions, which are also reflected in behavior.

Because the medical literature involving RT is so surprisingly vast, yet so specialized and heterogeneous, no attempt is made to review it in detail or to provide specific references. Those wishing to delve into these specialized uses of RT will find most of the literature referenced and abstracted, with an option to obtain the full articles in the Internet website http://www. MEDLINE. com/ (also medscape.com), by entering the key words with the name of the *medical condition of interest + reaction time* (e.g., Alzheimer disease + reaction time). Prominent among the highly diverse medical topics involving research with RT tests are:

> cognitive effects of normal aging, mild cognitive impairment, senile dementia, traumatic brain and closed head injuries, mortality, under-nutrition and malnutrition in children, eating disorders, parasitic infections, neurological effects of HIV and AIDS, drug effects and addictions, multiple sclerosis, sleep disorders, diabetes, attention deficit and hyperactivity disorder (ADHD), stroke, vascular dementia, degenerative brain diseases associated with aging (Huntington, Alzheimer, Parkinson), epilepsy, chronic fatigue syndrome, hypoxia, post-traumatic stress disorder (PTSD), psychiatric disorders (anxiety, schizophrenia, depression, bipolar), yoga and meditation, chemical, pharmaceutical, and nutriceutical (e.g., Gingko biloba) agents.

The most general conclusion that can be drawn from these studies is that in practically all of the above-mentioned conditions, the proband group when compared with a control group or a placebo control, showed statistically significant effects on RT and RTSD as measured by the diverse RT paradigms used in these studies.

It is hardly feasible to review this vast medical literature in the present book. RT itself seldom holds center stage in the overwhelming majority of these studies, as RT is usually included in a battery along with various psychometric tests, while the main focus is on the particular medical condition. Theoretical interpretations of the relationship between the particular medical condition and RT is typically nil or unsystematic. This does not imply that the study itself is not of considerable importance for reasons quite aside from its findings on RT.

After my perusal of over 100 recent articles in this medical literature, it appears that the main obstacle to a systematic review lies in the extreme heterogeneity of the chronometric methods used. Although they report statistically significant RT effects for a particular condition, some findings fail replication, or there are marked differences in effect size. However, true replications across different laboratories are seldom seen. The heterogeneity of studies includes a lack of uniformity in the paradigms, apparatuses, and procedures for measuring RT, such that method variance alone is inevitably huge, consisting of the direct effects of these varied testing conditions and also their statistical interactions with the great variability among studies in the age and sex of the probands. Another uncontrolled source of variance is the stage of development of the particular medical condition at the time RT is measured. The many combinations of all these variables are unique to virtually every study. Although the methodological uniqueness of a given study does not necessarily preclude its unique validity, it makes a meta-analysis highly problematic, if not impossible. This high degree of heterogeneity and methodological uniqueness of the vast majority of studies clearly hinders the development of a systematic and cumulative applied science of mental chronometry. Yet the development of such a systematic applied science could prove highly useful in medical and pharmaceutical research, and in monitoring the effects of treatment.

## Normative and Ipsative Applications

The distinction between normative and ipsative measurements is useful in discussing two broad classes of the medical uses of chronometry: (1) *diagnosis* of a condition and (2) *monitoring* the effects of treatment. Judging from the research literature so far, the diagnostic use of chronometry is the more problematic, particularly when it depends on normative data.

### Normative Differential Diagnosis

This is the comparison of an individual's performance on a test with the average performance of a *normative* (or control) group, which is usually a random sample of the general population, usually stratified by age. Even with ideally uniform methods and procedures for measuring RT, there is still the problem that the *mean* RT of any large *group* of patients with any kind of pathology that affects brain function will show a statistically significant difference from the normative group's mean, regardless of the particular RT paradigm. This phenomenon virtually rules out the comparison of a proband's RT with normative data as a discriminating tool for any differential diagnosis of an individual's specific condition, except possibly for probands that fall at the extremes of the population norms. Diagnostic precision might be moderately improved by stratifying the normative sample by age, IQ, education, occupation, and other major correlates of RT in a random population sample whose mean RT for a given paradigm reflects a multitude of variables that affect RT in general.

### Ipsative Diagnosis

This is based on the individual's pattern or *profile* of scores on a number of different tests, all measured on a common scale. The individual is his or her own "control." Although the overall average of the scores itself might be of diagnostic significance in its own right, the particular profile of high and low scores on the separate tests (e.g., different RT paradigms) offers the possibility of greater diagnostic specificity. But the kind of research that would be most informative regarding this possibility has not yet been done. First of all, it would call for a factor analysis in a normally diverse population, of a number of standard RT paradigms to determine the extent to which they reflect different latent factors besides the large general factor, or *g*, that appears to be common to all RT paradigms. To make this determination, it would require at least three forms of tests intended to measure the main information processing feature of each paradigm. For example, say we have three or more different forms or stimulus modalities of (a) the inspection time (IT) paradigm for measuring stimulus intake speed, (b) the Hick paradigm measuring RT of stimulus apprehension (SRT) and choice (CRT), (c) the Sternberg paradigm measuring RT for retrieval of information in STM, (d) the Posner paradigm measuring RT for retrieval of information from long-term memory (LTM), and (e) one or more dual task paradigms that reflect working memory capacity. The modality or type of information content would differ within each form of the RT paradigm. The question addressed to a factor analysis then is whether individual differences in each of these paradigms actually represent substantial uncorrelated factors, in addition to their one common factor. A nonsignificant or even very small factor for any particular paradigm would make it of questionable value as a discriminating diagnostic screening test in the general population. This paradigm, however, might still discriminate a pathological brain condition that is very scarce in the general population. For example, in a normal population, visual memory span and auditory memory span are perfectly correlated (after disattenuation), thus representing a single common factor. But the visual and auditory spans are almost totally uncorrelated in certain kinds of brain damage. Similarly, different RT paradigms that are not factor analytically distinguishable in a normal population could possibly show unique profile differences that would distinguish various brain pathologies. Some of the chronometric studies of Alzheimer disease and Parkinson disease, for example, note that Alzheimer mainly slows RT (decision time), reflecting cognitive impairment, whereas Parkinson mainly slows movement time (MT), reflecting impaired motor control.

### Criterion Analysis

One of the possible factor analytic methods for discovering the particular RT paradigms that may have unique diagnostic significance for a given well-defined and clearly identified condition is known as *criterion analysis* (Eysenck, 1950). It requires a normal *control* (C) group and a *proband* (P) group that has been diagnosed for a given condition as accurately as possible by the best means currently available. A factor analysis is performed on a battery of diagnostic tests (both chronometric and psychometric), including the point-biserial correlations of the C and P groups (quantized as 0 and 1) with each of the variables. An unrotated principal factor analysis reveals the location of the diagnostic criterion in the factor space, indicating the factors and tests most closely associated with the C–P distinction. Rotating the factor axes so that one of them passes directly through the C–P criterion further identifies the particular test variables that most strongly reflect the diagnostic criterion.

### Reaction Time and Movement Time

The RT literature is made problematic by the inconsistency across studies to employ methods that distinguish between the cognitive decision and the motor components of the task, treating them as if they are equivalent or indistinguishable. This occurs for all methods that measure only a single overt response to each presentation of the reaction stimulus (RS), thereby combining two different time components in the single measurement. But the decision and motor components clearly represent different latent traits, so combining them into a single measure will attenuate their distinctive significance. Three distinct time components are empirically measurable by the method that, on the occurrence of the RS, requires the subject to (A) release a home button which is the same for all tasks and (B) to press a selected response button. The total measured time for A when it is followed by B is called the RT or decision time (DT). The interval between A and B is the MT. But A actually consists of two distinguishable time components: (A1) the time for perception and encoding of the RS and the cognitive information processing operations involved in making the correct discrimination and choice response and (A2) preparation time or programing for the specific motor response called for by the RS. Thus RT = A1+A2. The time interval A2 is measured by requiring the subject to release the home button on the appearance of the RS, but not to press any response buttons (all of which may be completely covered). The non-motor or strictly cognitive information processing component of RT, then, is measured by subtraction: A1 = RT–A2. In the Hick paradigm, at least, the motor preparation component of RT is typically quite small compared to the cognitive component, so it is usually not determined (Jensen, 1987a). Theoretically, however, it would be valuable to know just how factor analytically distinct the motor preparation time is from the strictly cognitive processing time.

RT and MT are factor analytically distinct factors. Also, while RT shows large mean variations as a function of differences in the cognitive complexity of the purely information processing demands of the task, MT reflects this aspect only if the cognitive demands are also associated with considerably greater perceptual-motor demands on the response panel *per se*. The slight but significant correlation ($r<.30$) often found between RT and MT has not yet been thoroughly analyzed, but their small common factor does not seem to be *g* but rather some task specific factor not even identifiable as a true factor common to different RT–MT paradigms. The low correlation of MT with IQ, which unlike RT does not increase as a function of task complexity, could well be what has been termed an *extrinsic* correlation — a correlation without any functional connection between the two extrinsically correlated variables (Jensen, 1998b, pp. 139–143). The correlation between height and IQ is an example of extrinsic correlation, as shown by the fact that although these variables are correlated about + .20 in the general population, there is zero correlation between height and IQ among full siblings, which proves that there is no functional relationship between height and IQ. The correlation between MT and IQ could be of the same nature; but this hypothesis has not yet been tested. It could be most rigorously tested in a reasonably large sample of dizygotic twins, which would obviate any statistical corrections for age differences as would be required if ordinary siblings were used. (The genetic logic and analytic methodology for this kind of study are explained in the above reference, p. 141.)

# Monitoring Cognitive Change

The practical application in health care for which chronometry is probably most ideally suited is monitoring temporal changes in cognitive functions in (a) the normal course of aging, (b) the progression of some brain pathology that affects cognition, and (c) the response to treatment by drugs, physiotherapy, or other remedial interventions. Also, many drugs that are prescribed for ailments other than a cognitive condition can also have a sufficiently significant effect on cognition to call for monitoring, especially if the effect is related to activities of importance in the normal course of the patient's life, such as work efficiency, persistence, and attentional vigilance.

Compared to conventional psychometric tests, chronometry is especially well suited for all these purposes because of its high *sensitivity* for detecting variation in cognitive efficiency and its virtually unlimited *repeatability*.

The *sensitivity* of RT is attested by the fact that it is affected by small doses of stimulants or depressive drugs (e.g., caffeine and alcohol), normal diurnal metabolic fluctuations in body temperature, and the effects of physical (aerobic) exercise. Also, some RT measures have been shown to reflect early signs of mild cognitive impairment even before this condition can be detected by clinical assessment using conventional psychometric tests of memory, such as the memory test.

The *repeatability* of RT tests is probably their greatest advantage in monitoring cognitive efficiency. One and the same RT test (or a virtually unlimited number of practically exact equivalent forms) can be administered repeatedly, unlike the conventional tests based on items calling for specific knowledge or skills or the solution to problems that can be retained in the subject's memory or can be improved upon from one administration of the test to each subsequent administration.

## *Establishing an Individual's Reaction Time Baseline*

Ideally, from a medical diagnostic standpoint, every individual's general physical examination would include a chronometric screening of several basic cognitive functions that can be measured efficiently with a set of computer-administered RT–MT paradigms. The interpretation of the results obtained on each subsequent physical exam depends, of course, on initially establishing the given individual's baseline performance on each of the RT tasks over a period of several days, preferably under drug-free conditions and at the same time each day. The sensitivity of the highly accurate chronometric measures (in milliseconds) results in their considerable fluctuations from day-to-day and even at different times of the same day. Therefore, after sufficient practice trials on a given paradigm have been administered until familiarization with the task is assured and the initially large practice effects have leveled off, the individual's normal day-to-day variability can be determined as the level of "noise" above which that individual's RT must differ for it to indicate statistically significant change. It is analogous to noting changes in a person's weight or blood pressure on each periodic exam, as these measures, too, normally fluctuate around that person's central tendency. Because there are individual differences in the person's normal temporal variability, it must be taken into account for each person. At present, we do not have satisfactory normative data on intraindividual differences in daily fluctuations in the RT obtained on the several classic RT paradigms, although such normative data for given age groups would be most useful. At present, however, such norms would hardly be worth obtaining unless truly standardized methods of chronometric measurement were generally adopted, though alas, this *desideratum* has not yet been realized — more than a century since Donders (1868/1969) introduced mental chronometry to psychology! The importance of achieving such standardization is the entreaty of the following chapter.

# Chapter 14

# Standardizing Chronometry

The history of science is replete with evidence that important theoretical advances in empirical disciplines often depend on the development of highly standardized methods for measuring the phenomena under investigation. Universal agreement among scientists on the standardized CGS (centimeters, grams, seconds) units of measurement was essential for the remarkable progress of the physical sciences in the nineteenth century. Standardized measurement will be similarly important for the advancement of the behavioral and brain sciences in the twenty-first century.

## Psychometric Standardization

In traditional psychometrics, test standardization has two main features: (1) administering the test follows a standard procedure with explicit instructions for testing individuals qualified to take the particular test in terms of age, ability, language, and educational background. (2) The test is standardized, or "normed," in a sample of some clearly defined population (the *normative sample*). The distribution of raw scores in the normative sample is the basis for deriving standardized scores such as percentile ranks or $Z$ scores (or some linear transformation of $Z$, such as IQ). The scores derived from this standardization are wholly unique to the particular test and the normative sample. Individuals who score in widely different ranges of the total distribution may not even have taken the same items because many are either too easy or too difficult to contribute reliable variance outside a very narrow range of ability. Both the raw scores and any standard scores derived from them constitute an ordinal (rank order) scale. Any monotonic transformation of these scores remains an ordinal scale. Score intervals throughout the range are not necessarily equal units of the variable being assessed, nor are the interscore intervals necessarily equivalent to those of any other specific test intended to measure the same variable. Rank scores on different tests, however, can be very highly correlated even if they are not numerically equivalent, but the only standardized or constant features of psychometric tests are the test materials and the explicit administration procedure.

## Chronometric Standardization

Here, both historically and at the present time, the only standardized feature, ideally, is that in all chronometric studies the units of measurement are standardized as seconds (usually registered in terms of milliseconds) in the international CGS system. The timing mechanisms of present-day chronometric apparatuses are presumably accurate within a margin of error at the most of a few milliseconds. These units of time of course constitute a true ratio scale, with all its noted advantages. At present, this is virtually all that is truly standardized in chronometric research.

## Method Variance

In examining the chronometric test apparatuses improvised by many psychological laboratories in America, Europe, and Asia, I have found diversity even among apparatuses intended to measure response times (RTs) for nominally the same elementary cognitive task (ECT), for instance, differences in the angle of the response console, the size of the push-buttons and the force they require, and the degree of stimulus and response compatibility. These seemingly small differences between apparatuses used for nominally the same ECT yield consistently different absolute values of the sample means and standard deviations (SDs) of RT and movement time (MT), even though the distinctive pattern of the statistics for a given ECT generally remains ordinally the same across different apparatuses.

The discrepancy in the absolute measurements of nominally the same variable constitutes *method variance*. We have little knowledge of its overall magnitude in the chronometric research literature, although it is probably considerable.

From a scientific standpoint, method variance is unacceptable. It is confounded with subject-sampling variance and also interacts with the experimentally manipulated variables in any given ECT, making it impossible to know how much of the variation between studies from different laboratories consists of subject-sampling differences and how much is due to variation in the RT apparatuses and testing procedures. Of course, this adulterates the unique precision and generality of chronometric ratio-scale measurement, which are especially consequential for differential psychology (Jensen, 2004b). There is enough uncontrollable error variance in psychological research without allowing this tractable source of method variance to persist.

The solution is to standardize the chronometric apparatus and the testing procedures. Establishing norms for a given population is not intrinsically essential for most chronometric research, yet it may serve special purposes such as studying population trends in cognitive abilities or assessing an individual's cognitive status among a normative sample of age peers.

The summum bonum of having standardized chronometric methods in differential research, however, is that statistical differences between studies of nominally the same chronometric variables exclude method variance as much as possible and represent only variance between subjects and between explicitly different experimental conditions. Measurements representing absolute values in mental chronometry then would be directly comparable across studies conducted in different laboratories. They would be equivalent to standardized measures of physical variables that are universally accepted in the natural sciences. In principle, the same subjects should show numerically identical true-score measurements when they are obtained in different laboratories. This is a much more demanding and scientifically valuable criterion for RT measurement than simply finding that the data and statistics obtained across various studies based on different apparatuses and procedures are only ordinally similar.

# Desiderata for a Standardized Chronometric Apparatus

It would be desirable, if possible, for an international consortium of chronometric researchers to agree on the important features to be incorporated in a uniform, standardized apparatus and procedure that are recommended for adoption by all laboratories engaged in chronometric research. To set this prospect in motion, it might be useful to outline the main methodological problems that should be considered. Granting that present proposals are arguable and many details remain to be worked out, I shall not hesitate to voice my opinions on the critical issues.

The aim of chronometric standardization incorporates two main features: (1) a uniform standard RT apparatus consisting of a *display screen* for presenting the response stimuli (RS) of various standardized ECTs, along with the subject's *response console* for registering RT accurately in milliseconds; and (2) a set of explicit *procedures* for obtaining chronometric data with this apparatus.

## Procedures

Explicit uniform instructions for taking the particular ECT would be given by the examiner. A set number of practice trials on the task would precede the test trials. When it is questionable that the subject has understood the task, instructions are repeated and additional practice trials are given and recorded. Practice trials should be typical of the RS items used in the test trials and should be presented (or repeated) in a different random sequence to prevent the subject's memorizing nonessential features of the task (e.g., the serial order of the RS). All practice trials are recorded as such.

The number of test trials to be given depends on the reliability of the summary measures of the subjects' RT performance required for testing the hypothesis of interest in the particular study. Many RT studies in the past were based on very limited samples, resulting in statistical type 2 error (i.e., not rejecting the null hypothesis when it is false).

Whichever ECT is used as the primary focus of the particular investigation, an essential element in describing the subject sample in addition to the usual descriptors such as age, sex, education, or other demographic variables, is the descriptive statistics on the study sample's RT performance on brief tests of simple and choice RT and MT. This information based on standardized apparatus and procedure serves as a primary gauge, or baseline, for anchoring a particular subject sample's SRT and CRT, allowing comparison with subject samples in other studies. This summary information should be routinely reported in the appendix of every published study.

## Apparatus

Detailed descriptions and specifications of apparatus are inappropriate in the present context, but a few conditions should be considered regarding the stimulus *display screen* and the subject's *response console*. These features are crucial for standardization.

The subject sits in front of the display screen in a secretarial chair (without arm rests) adjustable for height to ensure the subject's comfortable view of the screen when it is tilted at about a 45° angle in relation to the horizontally placed response console directly in front of the display screen.

The *display screen* involves the variables of (a) *size*, (b) background *color* and contrasting stimulus color, (c) *luminosity* of the displayed stimulus and the ambient lighting in the testing room, and (d) the screen's *refreshment time* for the displayed stimuli.

Comments on the above: (a) the screen should not be too large, preferably about 25 cm². (b) RT is relatively sensitive to variation in visual–perceptual conditions, so the screen background should preferably be either white, blue, or black, with a uniformly contrasting color of the displayed stimulus. The aim is to minimize the effect of individual differences in color discrimination *per se*. (c) The *luminosity* of the display should be uniformly standardized to an optimum level, which is yet to be determined. The same goes for the *ambient lighting*, which can be checked with a photometer and adjusted with a rheostat. (d) *Refreshment time* is the speed with which a stimulus completely vanishes from the display screen to be followed immediately by another stimulus in the same location. Minimal refreshment time is critical for tests of individual differences in perceptual intake speed such as the inspection time (IT) paradigm. The fastest refreshment time that can be achieved with today's most advanced computer technology is probably adequate, but to attain maximum precision in measuring individual differences or subtle systematic intraindividual fluctuations in IT a display screen using light-emitting diodes (LEDs) might still be necessary.

*Response consoles* are probably the most variable component among the apparatuses in current use. Each consists of one or a combination of several generic properties: there are three main types of response consoles.

**The standard computer keyboard**   This typically calls for a *press* rather than a *lift* response and seldom controls which hand or fingers the subject uses for making the response. The four cursor keys are most frequently used. Because these keys are on the right-hand side of the keyboard, they possibly favor right-handed subjects. MT is typically not measured independently of RT, thus losing response information of value in the data analysis. RT and MT should not be amalgamated as one measurement because RT and MT are distinct factors that interact differently with other cognitive and motor skills, and with age. Subjects differ markedly in their familiarity with the standard keyboard, a difference that cannot be overcome by a few practice trials. The standard keyboard generally fails to control irrelevant variation in the subject's method of response, which inflates the measurement error variance with an error component that is largely avoidable. Using more than one finger of one hand for either RT or MT also introduces avoidable error variance. An even more undesirable condition is to require responses to be made on a set one of two or more of the keys by poising different fingers of one or both hands over a designated set of keys. All fingers are not equal! Subjects also find it distracting or confusing to have the whole complex keyboard exposed when only a few keys are needed for the particular task. Of course, all of the unused elements on the keyboard can be masked out, but this makeshift is time wasting and inelegant. Many obvious disadvantages of using a computer keyboard as the response console make it the least desirable choice for minimizing irrelevant and unwanted sources of variance in chronometric measurements.

It is most desirable to keep the strictly response requirements of the various RT paradigms as simple and uniform as possible. Otherwise one confounds the task's stimulus complexity and response complexity, which have independent effects on RT and MT. Ideally, only one and the same response (preferably lifting the index finger of the preferred hand) on one and the same push-button is required for absolutely every paradigm or ECT, thereby allowing only variation in response latencies. This crucial condition requires a *specialized response console*.

**Specialized response console**   This simple binary response console could also be called a *general purpose console*, because so many different tasks and problems can be presented in a strictly binary fashion (see Figure 2.4). Designed for use with each and every RT paradigm in which the response can be expressed as a binary choice, its central and essential feature is the *home button* (HB). A trial on any given RT paradigm begins with the subject's index finger of the preferred hand lightly pressing down on the HB, which is centrally located equidistant below the *choice response buttons* (CRBs) on the response console. Then, following an auditory *preparatory signal* (PS), the *response stimulus* (RS) appears on the display screen; the subject evaluates the RS and responds by releasing the HB and pressing the particular response button (RB) on the console representing the appropriate choice response. These discrete events occurring within every trial create four distinctly measurable time intervals:

HB–PS (a constant 1 s).

PS–RS (called the *preparatory interval*, it is usually a continuous variable of 1–3 s randomized over trials.

RS–HB (after the occurrence of the RS, the subject releases the HB. This interval is the RT or *decision time* (DT)).

HB–CRB (after releasing the HB, the subject presses the CRB. This interval is the MT).

*Caveat.* The use of RT apparatuses constructed by different laboratories, even when they follow specifications intended to produce clones, is a doubtful means for insuring the benefits of standardized measurement. My experience warns me not to expect complex apparatuses improvised independently in different psychological laboratories to be equivalent and interchangeable in registering identical measurements of RT or MT for nominally the same information-processing tasks. The benefits of chronometric standardization would be best assured if every laboratory used an exactly uniform apparatus — display screen, response console, and computer programs — all produced by the same manufacturers.

## Standardized Elementary Cognitive Tasks

In addition to a standardized apparatus, it would also be advantageous to provide standardized computer programs for a number of classical paradigms, which were originally intended to measure the speed of various information processes. A number of such paradigms are described in Chapter 2. Responses to most of them can be accommodated by a single binary choice response console consisting of an HB and two RBs. The binary choices to be made in response to a given paradigm (such as yes–no, true–false, same–different, etc.) can be readily changed by placing magnetized labels adjacent to the particular CRBs.

Specific paradigm programs are not hardwired in the apparatus itself. Computerized programs are inserted on easily changeable diskettes. The computer-controlled RT apparatus would require relatively little storage capacity, because each subject's RT, MT, and errors on all practice and test trials are immediately recorded on a diskette for later analysis by any computer that accommodates the commercially available statistical programs.

The RT tasks, most appropriate for standardization at present are the classic paradigms; most of them are described in Chapter 2 (also see subject index). These tests are intended to measure perceptual intake speed (IT); simple and choice RT; RT as a function of information load measured in *bits* (the Hick paradigm); simple discrimination RT (odd-man-out); scanning speed in short- and long-term memory (the Sternberg and Posner paradigms); and the speed of simple visual scanning (Neisser paradigm); the dual-task paradigm, the semantic verification test (SVT), running-memory tests based on various classes of stimuli: verbal, numerical, abstract spatial, pictorial, and facial stimuli.

Also recommended is the development of a chronometric measure of vocabulary. Its rationale is that although the test words are relatively common, thereby minimizing response errors, the RTs can estimate the subject's total vocabulary. This chronometric spelling test consists of presenting singly on the display screen a succession of both single words and nonwords (paralogs), to which the subject responds by leaving the HB and pressing the *Yes* button or the *No* button to indicate whether or not the presented item is an actual word. Familiar, high-frequency words (e.g., classified as AA in the Thorndike–Lorge (1944) word count) are randomly interspersed with an equal number of paralogs, which look like words and are easily pronounceable letter combinations of the same lengths as the actual test words, but are not in the unabridged English dictionary or a dictionary of slang). RT, MT, and response errors are registered separately for words and paralogs.

To enable the discovery of latent traits in the chronometric realm, there should exist at least two or three distinct forms of essentially the same paradigm but differing in content, such as verbal, numerical, abstract spatial, and pictorial. Every paradigm should have a standardized procedure for administration, departures from which, of course, would disqualify it as a standardized measurement.

Each RT paradigm should also have a programmed option for *informative feedback* (visual or auditory) on response errors. Generally, however, in most RT testing informative feedback is omitted, as it often distracts and interferes with sustained attention, slowing the mean RT (RTm) and inflating the standard deviation of RT (RTSD). Informative feedback, however, is useful, for example, if the experimenter wishes to shift the subject's performance toward some targeted level of speed/accuracy trade-off.

The only standardized absolute constant in the whole arrangement (besides the administration procedure) is the RT apparatus itself — the display screen and response console. While it is most useful, even essential, to have standardized programs of the classic paradigms, the standardized RT apparatus would permit researchers to devise novel paradigms needed for testing a particular hypothesis. All new programs, of course, should be deposited in the office of the journal that published the research or in a special repository for chronometric programs and data, which would provide copies on request.

## Individual Statistical Summary Scores

Although the subject's specific raw data over all trials would be recorded on disc, it would be useful if the apparatus also delivered certain basic summary measures of an individual's overall performance in a given set of trials. The most informative summary statistics to be registered on disc and to be immediately printed out for the tester's inspection are: (1) the number ($n$) of test trials, (2) the error rate (number of errors/$n$); and, separately for both RT and MT: (3) the mean, (4) median, (5) SD, and (6) mean-squared successive differences (see Chapter 4, p. 67).

Other derived scores or complex-scoring formulas should be viewed with suspicion, as they may be purely ad hoc and without any theoretically interpretable metric. Some derived scores may also lose the advantage of ratio-scale units. Such derived scores are usually the result of attempting to combine too many performance parameters into a single composite score. Examples are seen in some scoring formulas that arithmetically combine measures of speed (RT) and accuracy (error rate). Because *errors* and *times* (RT or MT) are conceptually different dimensions with different units of measurement, attempting to combine them by any particular formula must either be theoretically justified or altogether eschewed. There are scientifically meaningful concepts, however, that combine two qualitatively different units of measurement, such as measuring the acceleration of gravity in terms of m/s/s. Decisions concerning the treatment of speed and accuracy of responses depend mainly on the hypothesis or theoretical model of the behavior under consideration. When both speed and response accuracy are key variables, it is usually best to measure the basic RT parameters separately for correct responses and error responses. (The treatment of response errors is discussed in Chapter 4, pp. 69–70.)

### *Combining Time Measures*

Suppose that we have measures of RT or MT obtained on a battery of several experimentally independent tests, and we want to combine them all as a single score that preserves the advantages of ratio scale and the units of the original measurements. There are three proper ways for doing this:

(1)  Calculate the simple *arithmetic mean* of the individual RT measures on all of the tests (only provided that all the raw measures are in the same absolute metric units, e.g., milliseconds). With ordinary psychometric test scores, however, the raw scores on every test, which have numerically arbitrary means and *SDs*, must all be transformed to standardized scores having the same mean and SD (e.g., *Z scores*).

(2)  Perform a *multiple regression* analysis, provided there is a reliable criterion that we want to predict from scores on a number of different tests. The regression analysis determines the optimum weights for each of the predictors when they are combined in a weighted composite score that achieves the largest possible multiple correlation with the criterion variable that can be obtained with the given set of predictors.

(3)  If there is no particular external criterion variable that we wish to predict, we can derive a weighted composite score that best represents the whole test battery in terms of having the maximum variance obtainable with the given set of tests. This is a *principal components* analysis of the $n$ test variables. It must be performed on the *variance–covariance* matrix (*not* on the correlation matrix) if the ratio-scale units of measurement are to be preserved in the principal component scores. The first principal component (PC1) is the weighting factor that yields unidimensional composite

scores with the largest variance that can be derived from a given battery of tests. Only by factoring the *covariance* matrix (with the test variances in the main diagonal) instead of the *correlation* matrix (which is simply the Z-standardized covariance matrix), is the ratio-scale advantage of chronometry fully realized. A components analysis performed on correlations, however, converts the true zero point of the measurement scale to some arbitrary and unknown value, in effect degrading the component scores from a ratio scale to an ordinal scale (see Chapter 8, pp. 140–146 on chronometric factor scores.)

## Toward a Science of Mental Chronometry

Chronometry is a uniquely valuable instrument for measuring interindividual and intraindividual variation in many cognitive phenomena. Even though the basic time measurements obtained in laboratories worldwide are accurately scaled in identical units (viz., milliseconds), the unique apparatuses and procedures used in different laboratories systematically infest the data with unidentified and unwanted method variance. Without standardized measurement across all chronometric laboratories, we lose the vital benefit of a truly cumulative science. For this reason, the potential advantage of chronometry's usefulness as an investigative tool in the behavioral and brain sciences has not been fully realized. Its promise for research in the behavioral and brain sciences, however, is apparent in the recent uses of chronometry in laboratories here and abroad.

It is wrong, however, to conceive of chronometry as just an adjunct to psychometry, or as a substitute. Psychometry, which is now a highly developed quantitative and statistical technology, is commonly regarded as perhaps the most impressive achievement of modern applied psychology. The time/cost efficiency and practical validity of psychometric tests amply prove their utility. Hence, there is no need for chronometry to assume most of the typical purposes of psychometric tests. Neither should psychometrics be regarded as a higher court to which chronometry must appeal for scientific status.

It is a fact that certain chronometric measures are correlated with various external psychological and physiological variables and therefore may provide a precision tool for their absolute scale measurement and detailed process analysis. But we should recognize that chronometry can also stand by itself, revealing a natural domain of scientific interest in its own right, calling for empirical analysis and explanation. It is obviously important to understand the intrinsic natural properties of individual variation in the different chronometric measurements themselves, regardless of whatever degree of correlation they might have with our psychometric tests. Information gained from the investigation of chronometric measures as such seems essential for their usefulness in scientific research on psychological variation as well as in their practical applications, such as the diagnosis and monitoring of treatments in medical conditions that affect brain functions.

Here are a few examples of the intrinsic questions concerning the most frequently used chronometric variables that presently loom in need of definitive answers that can be attained only through specifically aimed studies.

The nature of the relationship between RT and MT is especially puzzling. Despite the uniformly high reliability of both RT and MT, the RT–MT correlations in different ECTs are highly erratic, ranging from .00 to .30 or .40. In some ECTs, such as the SVT, RTm shows a strong relationship to task complexity, whereas MT does not vary in the least with task complexity. RT–MT correlations also range widely with age, the scatter diagram showing a U-shaped function across age groups from childhood to the elderly. Since RT itself necessarily involves some degree of motor response as well as cognitive processing, is RT contaminated to varying degrees with the same motor aspect that is measured by MT? Or does the cognitive aspect of RT "spill over" into the measurement of MT in subjects who solve the ECT in mid-air, so to speak, while moving a hand from the HB to the selected RB? Or does RT simply reflect Fitts's law, which states that the RT to the subsequent performance of a motor task differs as a function of the complexity of the task? But then it is puzzling to find that increasing the number of response alternatives in the Hick paradigm, which increases the task's perceptual–motor demands, shows a strong relationship to RT, but shows no relationship whatever to MT. By completely omitting the MT response in the Hick paradigm, Hick's law still holds perfectly for RT; that is, the overall RTs are slower for the larger number of available response alternatives. In large-scale factor analyses, RT and MT load on uncorrelated factors, and in relation to psychometric factors RT has strong loadings on *g* as contrasted with the virtually zero *g* loadings of MT, which loads significantly only on a factor that exclusively represents MT as measured in various ECTs.

Yet, significant first-order correlations between RT and MT of around .30 are also found in many studies. Are these RT–MT correlations merely *extrinsic*, like the well-established correlation of about .20 between height and IQ, although there is absolutely no functional relationship between these two variables? If there were a functional relationship between RT and MT, it should show up in a large sample of dizygotic twins. One member of each twin pair would, on average, consistently score higher than his or her cotwin on *both* RT and MT. In other words, there would be a significant within-family correlation between RT and MT. In the absence of a within-family correlation, the population correlation between RT and MT would simply represent genetic heterogeneity in the population along with some common assortment of the particular genes that separately affect RT and MT. If this were indeed the case, it would further underline the importance of measuring RT and MT as distinctly as possible, and the measurement of MT, which appears to be a wholly noncognitive variable, would still be most useful for statistically removing or minimizing the purely motor contaminants from the RT measurements of cognitive speed. This should not imply that MT might not have other important correlates outside of the cognitive sphere. These are yet to be discovered. What now is most needed is the secure establishment of these seemingly inconsistent relationships between RT and MT and the formulation of a unified empirically testable theory that can accommodate them all as well as predict as yet undiscovered phenomena involving RT and MT. Scientific progress, in part, is a battle against the proliferation of narrowly specific ad hoc theories concocted to explain each and every newly observed phenomenon, like having one theory to explain why a thermos keeps liquids hot and another theory to explain why a thermos keeps liquids cold.

Another intrinsic question for theoretical analysis concerns the perfect disattenuated correlation between individual differences in the RTm and the RTSD over *n* trials. What causes this correlation? Is either of these variables causally primary — the speed of RT *per se* or the intertrial consistency of RT? This question quickly leads to questions about the neu-rophysiological basis of mental speed. The answer is still speculative. Is it nerve-conduc-tion velocity (NCV), which is known to be related to the degree of myelination of axons constituting the brain's white matter? Or is the causal factor the total number of neurons involved in information processing, so that the greater the number of neural pathways and their dendritic arborization involved in any particular cognitive task, the greater would be the reliability or consistency of the neural impulses that lead to evoking a correct response? Are the neural impulses cyclical, with individual differences in the period of the cycles such that faster cycling speed results both in faster RT and smaller RTSD? If the action potentials for response evocation are consistently cyclic and the cycle time differs between individuals, it could explain the reliable individual differences in RTSD. RTSD could also reflect a purely random intertrial fluctuation of potential but with consistent individual dif-ferences determining its maximal–minimal limits.

These are merely two examples of the kinds of basic questions that are intrinsic to developing a science of mental chronometry. Further examples can be gleaned from pre-vious chapters. But one hesitates to urge further basic chronometric research without first assuring the benefits of exact replication and generalizability of the results afforded by a standardized apparatus and procedure. Given such standardized conditions, however, chronometry provides the behavioral and brain sciences with a universal absolute scale for obtaining highly sensitive and frequently repeatable measurements of an individual's per-formance on specially devised cognitive tasks. Its time has come.

Let us get to work!

# References

Abeles, M. (2004). Time is precious. *Science, 304*, 523–524.

Ackerman, P. L., Beier, M. E., & Boyle, M. O. (2005). Working memory and intelligence: The same or different constructs? *Psychological Bulletin, 131*, 30–60.

Ahern, S., & Beatty, J. (1979). Pupillary responses during information processing vary with Scholastic Aptitude Test scores. *Science, 205*, 1289–1292.

Alderton, D. L., & Larson, G. E. (1994). Cross-task consistency in strategy use and the relationship with intelligence. *Intelligence, 18*, 47–76.

Ananda, S. M. (1985). *Speed of information processing and psychometric abilities in later adulthood*. Unpublished doctoral dissertation. University of California, Berkeley.

Anderson, J. R. (1983). Retrieval of information from long-term memory. *Science, 220*, 25–30.

Anderson, M. (1988). Inspection time, information processing and the development of intelligence. *British Journal of Developmental Psychology, 6*, 43–57.

Anderson, M. (1992). *Intelligence and development: A cognitive theory*. Oxford: Blackwell.

Anderson, M., Reid, C., & Nelson, J. (2001). Developmental changes in inspection time: What a difference a year makes. *Intelligence, 29*, 475–486.

Baddeley, A. D. (1968). A three-minute reasoning test based on grammatical transformation. *Psychonomic Science, 10*, 341–342.

Baddeley, A. D. (1986). *Working memory*. New York: Oxford University Press.

Baddeley, A. D., Thompson, N., & Buchanan, N. (1975). Word length and the structure of short-term memory. *Journal of Verbal Learning and Behavior, 14*, 575–589.

Baker, L. A., Vernon, P. A., & Ho, H.-Z. (1991). The genetic correlation between intelligence and speed of information processing. *Behavior Genetics, 21*, 351–367.

Barrett, L. F., Tugade, M. M., & Engle, R. W. (2004). Individual differences in working memory capacity and dual process theories of mind. *Psychological Bulletin, 130*, 553–573.

Barrett, P. T., Petrides, K. V., & Eysenck, H. J. (1998). Estimating inspection time: Response probabilities, the BRAT algorithm, and IQ correlations. *Personality and Individual Differences, 24*, 405–419.

Bartholomew, D. J. (2004). *Measuring intelligence: Facts and fallacies*. New York: Cambridge University Press.

Bartlett, M. S. (1937). The statistical conception of mental factors. *British Journal of Psychology, 28*, 97–104.

Baumeister, A. A. (1968). Behavioral inadequacy and variability of performance. *American Journal of Mental Deficiency, 73*, 477–483.

Baumeister, A. A. (1998). Intelligence and the "personal equation." *Intelligence, 26*, 255–265.

Baumeister, A. A., & Kellas, G. A. (1967). Refractoriness in the reaction times of normals and retardates as a function of response-stimulus interval. *Journal of Experimental Psychology, 75*, 122–125.

Baumeister, A. A., & Kellas, G. (1968a). Reaction time and mental retardation. In: N. R. Ellis (Ed.), *International review of research in mental retardation*, Vol. 3, pp. 163–193. New York: Academic Press.

Baumeister, A. A., & Kellas, G. (1968b). Distribution of reaction times of retardates and normals. *American Journal of Mental Deficiency, 72*, 715–718.

Baumeister, A. A., & Maisto, A. A. (1977). Memory scanning by children: Meaningfulness and mediation. *Journal of Experimental Child Psychology, 24*, 97–107.

Baumeister, A., Urquhart, D., Beedle, R., & Smith, T. (1964). Reaction times of normals and retardates under different stimulus intensity changes. *American Journal of Mental Deficiency, 69*, 126–130.

Baumeister, A. A., & Ward III, L. C. (1967). Effects of rewards upon the reaction times of mental defectives. *American Journal of Mental Deficiency, 71*, 801–805.

Baumeister, A. A., & Wilcox, S. J. (1969a,b). Effects of variations in the preparatory interval on the reaction times of retardates and normals. *Journal of Abnormal Psychology, 74*, 438–442.

Beaujean, A. A. (2005). Heritability of cognitive abilities as measured by mental chronometric tasks: A meta-analysis. *Intelligence, 33*, 187–201.

Beck, L. F. (1933). The role of speed in intelligence. *Psychological Bulletin, 30*, 169–178.

Beier, M. E., & Ackerman, P. L. (2005). Working memory and intelligence: Different constructs. Reply to Oberauer et al. (2005) and Kane et al. (2005). *Psychological Bulletin, 131*, 72–75.

Bigler, E. D., Johnson, S. C., Jackson, C., & Blatter, D. D. (1995). Aging, brain size, and IQ. *Intelligence, 21*, 109–119.

Bódizs, R., Kis, T., Lázár, A. S., Havrán, L., Rigó, P., Clemens, Z., & Halász, P. (2005). Prediction of general mental ability based on neural oscillation measures of sleep. *Journal of Sleep Research, 14*, 285–292.

Boomsma, D. I., & Somsen, R. J. M. (1991). Reaction times measured in a choice reaction time and a double task condition: A small twin study. *Personality and Individual Differences, 12*, 519–522.

Boring, E. G. (1950). *A history of experimental psychology*. New York: Appleton-Century-Crofts.

Bors, D. A., & Forrin, B. (1995). Age, speed of information processing, recall, and fluid intelligence. *Intelligence, 20*, 229–248.

Bouchard, T. J., Jr., Lykken, D. T., Segal, N. L., & Wilcox, K. J. (1986). Developmental in twins reared apart: A test of the chronogenetic hypothesis. In: A. Demirjian (Ed.), *Human growth: A multidisciplinary review*. London: Taylor & Francis.

Brand, C., & Deary, I. J. (1982). Intelligence and 'inspection time. In: H. J. Eysenck (Ed.), *A model for intelligence*, pp. 133–148. Berlin: Springer.

Brewer, N., & Smith, G. A. (1989). Developmental changes in processing speed: Influence of speed-accuracy regulation. *Journal of Experimental Psychology: General, 118*, 298–310.

Brinley, J. F. (1965). Cognitive sets, speed and accuracy of performance in the elderly. In: A. T. Welford, & J. E. Birren (Eds), *Behavior, aging and the nervous system*, pp. 114–149. Springfield, IL: Charles C. Thomas.

Brody, N. (1992). *Intelligence*, 2nd ed. San Diego: Academic Press.

Bull, R., & Johnston, R. S. (1997). Children's arithmetical difficulties: Contributions from processing speed, item identification, and short-term memory. *Journal of Experimental Child Psychology, 65*, 1–24.

Burns, B. D. (2004). The effects of speed on skilled chess performance. *Psychological Science, 15*, 442–447.

Burns, N. R., & Nettelbeck, T. (2003). Inspection time in the structure of cognitive abilities: Where does IT fit? *Intelligence, 31*, 237–255.

Burrows, D., & Okada, R. (1975). Memory retrieval from long and short lists. *Science, 188*, 1031–1032.

Burt, C. (1940). *The factors of the mind*. London: University of London Press.

Buzsaki, G., & Draguhn, A. (2004). Neuronal oscillations in cortical networks. *Science, 304*, 1926–1929.

Carroll, J. B. (1987). Jensen's mental chronometry: Some comments and questions. In: S. Modgil, & C. Modgil (Eds), *Arthur Jensen: Consensus and controversy*, pp. 297–307. New York: Falmer.

Carroll, J. B. (1991a). Reaction time and cognitive abilities: Review of factor-analytic studies. Paper presented at a symposium, "reaction time and mental abilities, history, status and trends." Annual meeting of the American Psychological Association, San Francisco, CA, August 19, 1991.

Carroll, J. B. (1991b). No demonstration that *g* is not unitary, but there's more to the story: Comment on Kranzer and Jensen. *Intelligence, 15*, 423–436.

Carroll, J. B. (1993). *Human cognitive abilities: A survey of factor analytic studies*. Cambridge, UK: Cambridge University Press.

Caryl, P. G., Deary, I. J., Jensen, A. R., Neubauer, A. C., & Vickers, D. (1999). Information processing approaches to intelligence: Progress and prospects. In: I. Mervielde, I. Deary, F. de Fruyt, & F. Ostendorf (Eds), *Personality psychology in Europe*, Vol. 7. Tilburg: Tilburg University Press.

Case, R. (1985). *Intellectual development: Birth to adulthood*. Orlando, FL: Academic Press.

Cattell, J. M. (1929). Psychology in America. *Science, 70*, 335–347.

Cattell, J. McK. (1886). The time taken up by cerebral operations. *Mind, 11*, 220–242; 377–392.

Cavanagh, P. (1972). Relation between the immediate memory span and the mental search rate. *Psychological Review, 79*, 525–530.

Cerella, J. (1985). Information processing rates in the elderly. *Psychological Bulletin, 98*, 67–83.

Cerella, J., DiCara, R., Williams, D., & Bowles, N. (1986). Relations between information processing and intelligence in elderly adults. *Intelligence, 10*, 75–91.

Cerella, J., & Hale, S. (1994). The rise and fall in information-processing rates over the life span. *Acta Psychologica, 86*, 109–197.

Chaiken, S. R. (1994). The inspection time not studied: Processing speed ability unrelated to psychometric intelligence. *Intelligence, 19*, 295–316.

Cohn S. J., Carlson, J. S., & Jensen, A. R. (1985). Speed of information processing in academically gifted youths. *Personality and Individual Differences, 6*, 621–629.

Colom, R., Rebollo, I., Palacios, A., Juan-Espinosa, M.I., & Kyllonen, P. C. (2004). Working memory is (almost) perfectly predicted by *g*. *Intelligence, 32*, 277–296.

Conway, A. R. A., Kane, M. J., & Engle, R. W. (2000). Is Spearman's *g* determined by speed or working memory capacity? *Psycoloquy, 11*(038). Internet address: http://www.cogsci.soton.ac.uk/cgi/psyc/newpsy.

Cooper, L. A. (1975). Mental rotation of random two-dimensional forms. *Cognitive Psychology, 7*, 20–43.

Cooper, L. A., & Shepard, R. N. (1973). Mental rotation of letters. In: W. G. Chase (Ed.), *Visual information processing*. New York: Academic Press.

Couvée, J. E., Van den Bree, M. B. M., & Orlebeke, J. F. (1988). Genetic analysis of reaction time in twins and their parents. Paper presented at the 18th annual meeting of the Behavior Genetics Association, University of Nijmegen, The Netherlands, June 22–25.

Coyle, T. R. (2003). IQ, the worst performance rule, and Spearman's law: A reanalysis and extension. *Intelligence, 31*, 473–489.

Cronbach, L. J. (1957). The two disciplines of scientific psychology. *American Psychologist, 12*, 671–684.

Dalteg, A., Rasmussen, K., Jensen, J., Persson, B., Lindgren, M., Lundqvist, A., Wirsen-Meurling, A., Ingvar, D. H., & Levander, S. (1997). Prisoners use an inflexible strategy in a continuous performance test: A replication. *Personality and Individual Differences, 23*, 1–7.

Deary, I. J., (1993). Inspection time and WAIS-R IQ subtypes: a confirmatory factor analysis study. *Intelligence, 17*, 223–236.

Deary, I. J. (1994). Sensory discrimination and intelligence: Postmortem or resurrection? *American Journal of Psychology, 107*, 95–115.

Deary, I. J. (1996). Reductionism and intelligence: The case of inspection time. *Journal of Biosocial Science, 28*, 405–423.

Deary, I. J. (2000a). *Looking down on human intelligence: From psychometrics to the brain*. Oxford: Oxford University Press.

Deary, I. J. (2000b). Simple information processing and intelligence. In: R. J. Sternberg (Ed.), *Handbook of intelligence*, pp. 267–284. Cambridge: Cambridge University Press.

Deary, I. J. (2003). Reaction time and psychometric intelligence: Jensen's contributions. In: H. Nyborg (Ed.), *The scientific study of general intelligence: A tribute to Arthur R. Jensen*, pp. 53–75. Oxford: Pergamon.

Deary, I. J., & Crawford, J. R. (1998). A triarchic theory of Jensenism: Persistent conservative reductionism. *Intelligence, 26*, 73–282.

Deary, I. J., & Der, G. (2005a). Reaction time, age, and cognitive ability: Longitudinal findings from age 16 to 63 years in representative population samples. *Aging, Neuropsychology, and Cognition, 12*, 1–29.

Deary, I. J., & Der, G. (2005b). On RT and age of death. *Psychological Science, 16*, 64–69.

Deary, I. J., Der, G., & Ford, G. (2001). Reaction times and intelligence differences: A population-based cohort study. *Intelligence, 29*, 389–399.

Deary, I. J., Egan, V., Gibson, G. J., Austin, E., Brand, C. R., & Kellaghan, T. (1996). Intelligence and the differentiation hypothesis. *Intelligence, 23*, 105–132.

Deary, I. J., McCrimmon, R. J., & Bradshaw, J. (1997). Visual information processing and intelligence. *Intelligence, 24*, 461–479.

Deary, I. J., & Stough, C. (1996). Intelligence and inspection time: Achievements, prospects and problems. *American Psychologist, 51*, 599–608.

Dempster, F. N. (1981). Memory span: Sources of individual and developmental differences. *Psychological Bulletin, 89*, 63–100.

Der, G., & Deary, I. J. (2003). IQ, reaction time and the differentiation hypothesis. *Intelligence, 31*, 491–503.

Detterman, D. K. (1987). What does reaction time tell us about intelligence? In P. A. Vernon (Ed.), *Speed of information-processing and intelligence*, pp. 177–200. Norwood, NJ: Ablex.

Detterman, D. K., Mayer, J. D., Caruso, D. R., Legree, P. J., Conners, F. A., & Taylor, R. (1992). Assessment of basic cognitive abilities in relation to cognitive deficits. *American Journal on Mental Retardation, 97*, 251–286.

Donders, F. C. (1969). On the speed of mental processes. *Acta Psychologica, 30*, 412–431. (English translation of the original article published in 1868.)

Egan, V., & Deary, I. J. (1992). Are specific inspection time strategies prevented by concurrent tasks? *Intelligence, 16*, 151–167.

Elliott, R. (1970). Simple reaction time: Effects associated with age, preparatory interval, incentive shift, and mode of presentation. *Journal of Experimental Child Psychology, 9*, 86–107.

Elliott, R. (1972). Simple reaction time in children: Effects of incentive, incentive-shift and other training variables. *Journal of Experimental Child Psychology, 13*, 540–557.

Eysenck, H. J. (1950). Criterion analysis and application of the hypothetico-deductive method to factor analysis. *Psychological Review, 57*, 38–53.

Eysenck, H. J. (1967). Intelligence assessment: A theoretical and experimental approach. *British Journal of Educational Psychology, 37*, 81–98.

Eysenck, J. J. (1987). Intelligence and reaction time: The contribution of Arthur Jensen. In: S. Modgil, & C. Modgil (Eds), *Arthur Jensen: Consensus and controversy*, pp. 285–296. New York: Falmer.

Fairweather, H., & Hutt, S. J. (1978). On the rate of gain of information in children. *Journal of Experimental Child Psychology, 26*, 216–229.

Fitts, P. M. (1954). The information capacity of the human motor system in controlling the amplitude of movement. *Journal of Experimental Psychology, 47*, 381–391.

Frearson, W., & Eysenck, H. J. (1986). Intelligence, reaction time, and a "odd-man-out" RT paradigm. *Personality and Individual Differences, 7*, 807–817.

Fry, A. F., & Hale, S. (1996). Processing speed, working memory, and fluid intelligence: Evidence for a developmental cascade. *Psychological Science, 7*, 237–241.

Fry, A. F., & Hale, S. (2000). Relationships among processing speed, working memory, and fluid intelligence in children. *Biological Psychology, 54*, 1–34.

Galley, N., & Galley, L. (1999). Fixation durations and saccaid latencies as indicators of mental speed. In: I. Mervielde, I. Deary, F. de Fruyt, & F. Ostendorf (Eds), *Personality psychology in Europe*, Vol. 7. Tilburg: Tilburg University Press.

Galton, F. (1908). *Memories of my life*. London: Methuen.

Garlick, D. (2002). Understanding the nature of the general factor of intelligence: The role of individual differences in neural plasticity as an explanatory mechanism. *Psychological Review, 109*, 116–136.

Glaser, W. R., & Dolt, M. O. (1977). A functional model to localize the conflict underlying the Stroop phenomenon. *Psychological Research, 39*, 287–310.

Glimcher, P. W. (1999). Eye movements. In: M. J. Zigmond, F. E. Bloom, S. C. Landis, J. L. Roberts, & L. R. Squire (Eds), *Fundamental neuroscience*. San Diego: Academic Press.

Grudnik, J. L., & Kranzler, J. H. (2001). Meta-analysis of the relationship between intelligence and inspection time. *Intelligence, 29*, 523–535.

Guilford, J. P. (1954). *Psychometric methods* (2nd ed.). New York: McGraw-Hill.

Hale, S. (1990). A global developmental trend in cognitive processing speed. *Child Development, 61*, 653–663.

Hale, S., & Jansen, J. (1994). Global processing-time coefficients characterize individual and group differences in cognitive speed. *Psychological Science, 5*, 384–389.

Hart, B. I. (1942). Tabulation of the probabilities for the ratio of the mean square successive difference to the variance. *Annals of Mathematical Statistics, 13*, 207–214.

Hemmelgarn, T. E., & Kehle, T. J. (1984). The relationship between reaction time and intelligence in children. *School Psychology International, 5*, 77–84.

Hertzog, C., & Bleckley, M. K. (2001). Age differences in the structure of intelligence: Influences of information processing speed. *Intelligence, 29*, 191–217.

Hick, W. E. (1952). On the rate of gain of information. *Quarterly Journal of Experimental Psychology, 4*(1), 11–16.

Ho, H. -Z., Baker, L. A., & Decker, S. N. (1988). Covariation between intelligence and speed of cognitive processing: Genetic and environmental influences. *Behavior Genetics, 18*, 247–261.

Hull, C. L. (1943). *Principles of behavior*. New York: Appleton-Century.

Hunt, E., Lunneborg, C., & Lewis, J. (1975). What does it mean to be high verbal? *Cognitive Psychology, 7*, 194–227.

Ikegaya, Y., Aaron G., Cossart, R., Aronov, D., Lampl, I., Ferster, D., & Yuste, R. (2004). Synfire chains and cortical songs: Temporal modules of cortical activity. *Science, 304*, 559–564.

Jensen, A. R. (1965). Scoring the Stroop test. *Acta Psychologica, 24*, 398–408.

Jensen, A. R. (1971). Individual differences in visual and auditory memory. *Journal of Educational Psychology, 62*, 123–131.

Jensen, A. R. (1977). Cumulative deficit in IQ of blacks in the rural South. *Developmental Psychology, 13*, 184–191.

Jensen, A. R. (1980a). Uses of sibling data in educational and psychological research. *American Educational Research Journal, 17*, 153–170.

Jensen, A. R. (1980b). *Bias in mental testing*. New York: Free Press.

Jensen, A. R. (1982). Reaction time and psychometric g. pp. 93–132. In: H. J. Eysenck (Ed.), *A model for intelligence*. New York: Springer.

Jensen, A. R. (1983). Critical flicker frequency and intelligence. *Intelligence, 7*, 217–225.

Jensen, A. R. (1985). Methodological and statistical techniques for the chronometric study of mental abilities. In: C. R. Reynolds, & V. L. Willson (Eds), *Methodological and statistical advances in the study of individual differences*. New York: Plenum Press.

Jensen, A. R. (1987a). Individual differences in the Hick paradigm. In: P. A. Vernon (Ed.), *Speed of information processing and intelligence*, pp. 101–175, Norwood, NJ: Ablex.

Jensen, A. R. (1987b). Process differences and individual differences in some cognitive tasks. *Intelligence, 11*, 107–136.

Jensen, A. R. (1990). Speed of information processing in a calculating prodigy. *Intelligence, 14*, 259–274.

Jensen, A. R. (1992). The importance of intraindividual variability in reaction time. *Personality and Individual Differences, 13*, 869–882.

Jensen, A. R. (1994). Francis Galton. In: R. J. Sternberg (Ed.), *Encyclopedia of intelligence*, Vol. 1, pp. 457–463. New York: Macmillan.

Jensen, A. R. (1996). Giftedness and genius: Crucial differences. In: C. P. Benbow, & D. Lubinski (Eds), *Intellectual talent: Psychometric and social issues*, pp. 393–411. Baltimore: Johns Hopkins University Press.

Jensen, A. R. (1998a). The suppressed relationship between IQ and the reaction time slope parameter of the Hick function. *Intelligence, 26*, 43–52.

Jensen, A. R. (1998b). *The g factor*. Westport, CT: Praeger.

Jensen, A. R. (2000). Charles E. Spearman: The discoverer of *g*. In: G. A. Kimble, & M. Wetheimer (Eds), *Portraits of pioneers in psychology*, Vol. IV. Washington, DC: American Psychological Association.

Jensen, A. R. (2002). Galton's legacy to research on intelligence. *Journal of Biosocial Science, 34*, 145–172.

Jensen, A. R. (2004a). The mental chronometry of giftedness. In: D. Boothe, & J. C. Stanley (Eds), *In the eyes of the beholder: Cultural and disciplinary perspectives in giftedness*, pp. 157–166. Waco, TX: Prufrock Press Inc.

Jensen, A. R. (2004b). Mental chronometry and the unification of differential psychology. In: R. J. Sternberg, & J. E. Pretz (Eds), *Cognition and intelligence*, pp. 26–50. New York: Cambridge University Press.

Jensen, A. R., Cohn, S. J., & Cohn, C. M. G. (1989). Speed of information processing in academically gifted youths and their siblings. *Personality and Individual Differences, 10*, 29–34.

Jensen, A. R., Larson, J., & Paul, S. M. (1988). Psychometric g and mental processing speed on a semantic verification test. *Personality and Individual Differences, 9*, 243–255.

Jensen, A. R., & Munro, E. (1979). Reaction time, movement time, and intelligence. *Intelligence, 3*, 121–126.

Jensen, A. R., & Reed, T. E. (1990). Simple reaction time as a suppressor variable in the chronometric study of intelligence. *Intelligence, 14*, 375–388.

Jensen, A. R., & Rohwer, W. D., Jr. (1966). The Stroop color-word test: A review. *Acta Psychologica, 25*, 36–93.

Jensen, A. R., Schafer, E. W. P., & Crinella, F. (1981). Reaction time, evoked brain potentials, and psychometric g in the severely retarded. *Intelligence, 5*, 179–197.

Jensen, A. R., & Sinha, S. N. (1993). Physical correlates of human intelligence. In: P. A. Vernon (Ed.), *Biological approaches to the study of human intelligence*, pp. 139–242. Norwood, NJ: Ablex.

Jensen, A. R., & Vernon, P. A. (1986). Jensen's reaction time studies: A reply to Longstreth. *Intelligence, 10*, 153–179.

Jensen, A. R., & Weng, L.-J. (1994). What is a good *g*? *Intelligence, 18*, 231–258.

Jensen, A. R., & Whang, P. A. (1993). Reaction times and intelligence: A comparison of Chinese-American and Anglo-American children. *Journal of Biosocial Science, 25*, 397–410.

Johnson, R. C., McClearn, G. E., Yuen, S., Nagoshi, C. T., Ahern, F. M., & Cole, R. E. (1985). Galton's data a century later. *American Psychologist, 40*, 875–892.

Kail, R. (1991a). Developmental change in speed of processing during childhood and adolescence. *Psychological Bulletin, 109,* 490–501.

Kail, R. (1991b). Development of processing speed in childhood and adolescence. *Advances in Child Development and Behavior, 23,* 151–185.

Kail, R. (1992). General slowing of information-processing by persons with mental retardation. *American Journal on Mental Retardation, 97,* 333–341.

Kail, R., & Park, Y. (1994). Processing time, articulation time, and memory span. *Journal of Experimental Child Psychology, 57,* 281–291.

Kane, M. J., Hambrick, D. Z., & Conway, A. R. A. (2005). Working memory capacity and fluid intelligence are strongly related constructs: Comment on Ackerman, Beier and Boyle. *Psychological Bulletin, 131,* 66–71.

Kendall, M. G., & Stuart, A. (1973). *The advanced theory of statistics,* 3rd ed., Vol. 2. London: Griffin.

Kendall, M. G., & Stuart, A. (1977). *The advanced theory of statistics,* 4th ed., Vol. 1. London: Griffin.

Kirby, N. (1980). Sequential effects in choice reaction time. In: A. T. Welford (Ed.), *Reaction times,* pp. 129–172. London: Academic Press.

Kline, P., Draycott, S. G., & McAndrew, V. M. (1994). Reconstructing intelligence: A factor analytic study of the BIP. *Personality and Individual Differences, 16,* 529–536.

Koch, C., Gobell, J., & Roid, G. H. (1999). Exploring individual differences in Stroop processing with cluster analysis. *Psycoloquy, 10*(025), Stroop Differences (1).

Kranzler, J. H. (1991). Reacton time, response errors, and *g. Perceptual and Motor Skills, 73,* 1035–1043.

Kranzler, J. H. (1992). A test of Larson and Alderton's (1990) worst performance rule of reaction time variability. *Personality and Individual Differences, 13,* 255–261.

Kranzler, J. H., & Jensen, A. R. (1989). Inspection time and intelligence: A meta-analysis. *Intelligence, 13,* 329–347.

Kranzler, J. H., & Jensen, A. R. (1991). The nature of psychometric *g*: Unitary process or a number of independent processes? *Intelligence, 15,* 397–422.

Kranzler, J. H., Whang, P. A., & Jensen, A. R. (1988). Jensen's use of the Hick paradigm: Visual attention and order effects. *Intelligence, 12,* 371–391.

Kranzler, J. H., Whang, P. A., & Jensen, A. R. (1994). Task complexity and the speed and efficiency of elemental information processing: Another look at the nature of intellectual giftedness. *Contemporary Educational Psychology, 19,* 447–459.

Kyllonen, P. C. (1985). *Dimensions of information processing speed.* Brooks Air Force Base, TX: Air Force Systems Command, AFHRL-TP-84-56.

Kyllonen, P. C. (1994). CAM: A theoretical framework for cognitive abilities measurement. In: D. Detterman (Ed.), *Current topics in human intelligence: Volume IV, theories of intelligence,* pp. 307–359. Norwood, NJ: Ablex.

Lally, M., & Nettelbeck, T. (1977). Intelligence, reaction time, and inspection time. *American Journal of Mental Deficiency, 82,* 273–281.

Laming, D. (1988). Some boundary conditions of choice reaction performance. In: I. Hindmarch, B. Aufdembrinke, & H. Ott (Eds), *Psychopharmacology and reaction time.* New York: Wiley.

Landis, C. (1953). *An annotated bibliography of flicker fusion phenomena, 1740–1952.* Ann Arbor: University of Michigan.

Larson, G. E. (1989). A brief note on coincidence timing. *Intelligence, 13,* 361–369.

Larson, G. E., & Alderton, D. L. (1990). Reaction time variability and intelligence: A "worst performance" analysis of individual differences. *Intelligence, 14,* 309–325.

Larson, G. E., & Saccuzzo, D. P. (1986). Jensen's reaction time experiments: Another look. *Intelligence, 10,* 231–238.

Larson, G. E., & Saccuzzo, D. P. (1989). Cognitive correlates of general intelligence: Toward a process theory of *g. Intelligence, 13,* 5–31.

Larson, G. E., Saccuzzo, D. P., & Brown, J. (1994). Motivation: Cause or confound in information processing/intelligence correlation? *Acta Psychologica, 85,* 25–37.

Lehrl, S., & Fischer, B. (1988). The basic parameters of human information process: Their role in the determination of intelligence. *Personality and Individual Differences, 9,* 883–896.

Lehrl, S., & Fischer, B. (1990). A basic information psychological parameter (BIP) for the reconstructions of concepts of intelligence. *European Journal of Personality, 4,* 259–286.

Leiderman, P. H., & Shapiro, D. (1962). Application of a time series statistic to physiology and psychology. *Science, 138,* 141–142.

Lemmon, V. W. (1927). The relation of reaction time to measures of intelligence, memory, and learning. *Archives of Psychology, 15,* 5–38.

Levine, G., Predy, D., & Thorndike, R. L. (1987). Speed of information processing and level of cognitive ability. *Personality and Individual Differences, 8,* 599–607.

Libet, B. (1989). Conscious subjective experience vs. unconscious mental functions: A theory of the cerebral processes involved. In: R. M. J. Cotterill (Ed.), *Models of brain function.* Cambridge: Cambridge University Press.

Li, S.-C., & Lindenberger, U. (1999). Cross-level unification: A computational exploration of the link between deterioration of neurotransmitter systems and dedifferentiation of cognitive abilities in old age. In: G.-G. Nilsson, & H. J. Markowitsch (Eds), *Cognitive neuroscience of memory,* pp. 103–146. Seattle: Hogrefe & Huber.

Liebert, A. M., & Baumeister, A. A. (1973). Behavioral variability among retardates, children, and college students. *The Journal of Psychology, 83,* 57–65.

Lindley, R. H., Smith, W. R., & Thomas, T. J. (1988). The relationship between speed of information processing as measured by timed paper-and-pencil tests and psychometric intelligence. *Intelligence, 12,* 17–26.

Lindley, R. H., Wilson, S. M., Smith, W. R., & Bathurst, K. (1995). Reaction time (RT) and IQ: shape of the task complexity function. *Personality and Individual Differences, 18,* 339–345.

Loehlin, J. C. (1998). *Latent variable models: An introduction to factor, path, and structural analysis* (3rd ed.). Mahwah, NJ: Erlbaum.

Lohman, D. F. (2000). Complex information processing and intelligence. In: R. J. Sternberg (Ed.), *Handbook of intelligence.* Cambridge, UK: Cambridge University Press.

Longstreth, L. E. (1984). Jensen's reaction-time investigations of intelligence: A critique. *Intelligence, 8,* 139–160.

Lubinski, D., & Humphreys, L. G. (1996). Seeing the forest from the trees: When predicting the behavior or status of groups, correlate means. *Psychology, Public Policy, and Law, 2,* 363–376.

Luce, R. D. (1986). *Response times: Their role in inferring elementary mental organization* (Oxford Psychology Series No. 8.). New York: Oxford University Press.

Luciano, M., Psthuma, D., Wright, M. J., de Geus, Eco, J. C., Smith, G. A., Geffen, G. M., Boomsma, D. I., & Martin, N. G. (2005). Perceptual speed does not cause intelligence and intelligence does not cause perceptual speed. *Biological Psychology, 70,* 1–8.

Luciano, M., Wright, M. J., Smith, G. A., Geffen, G. M., Geffen, L. B., & Marytin, N. G. (2001). Genetic covariance among measures of information processing speed, working memory, and IQ. *Behavior Genetics, 311,* 581–592.

Luo, D., & Petrill, S. A. (1999). Elementary cognitive tasks and their roles in *g* estimates. *Intelligence, 27,* 157–174.

Luo, D., Thompson, L. A., & Detterman, D. K. (2003). The causal factor underlying the correlation between psychometric *g* and scholastic performance. *Intelligence, 31,* 67–83.

Luo, D., Thompson, L. A., & Detterman D. K. (2006). The criterion validity of tasks of basic cognitive processes. *Intelligence, 34*, 79–120.

Mackenzie, B., Molley, E., Martin, F., Lovegrove, W., & McNicol, D. (1991). Inspection time and the content of simple tasks: A framework for research on speed of information processing. *Australian Journal of Psychology, 43*, 37–43.

Maisto, A. A., & Baumeister, A. A. (1975). A developmental study of choice reaction time: The effect of two forms of stimulus degradation on encoding. *Journal of Experimental Child Psychology, 20*, 456–464.

Mathews, G., & Dorn, L. (1989). IQ and choice RT: An information processing analysis. *Intelligence, 13*, 299–317.

McGaugh, J. L. (1966). Time-dependent processes in memory storage. *Science, 153*, 1351–1358.

McGue, M., & Bouchard, T. J., Jr. (1989). Genetic and environmental determinants of information processing and special mental abilities: A twin analysis. In: R. J. Sternberg (Ed.), *Advances in the psychology of human intelligence*, Vol. 5, pp. 7–45. Hillsdale, NJ: Erlbaum.

McGue, M., Bouchard, T. J., Jr. Lykken, D. T., & Feier, D. (1984). Information processing abilities in twins reared apart. *Intelligence, 8*, 239–258.

McNicol, D., & Stewart, G. W. (1980). Reaction time and the study of memory. In: A. T. Welford (Ed.), *Reaction times*, pp. 253–307. New York: Academic Press, Inc.

Merkel, J. (1885). Die zeitlichen Verhältnisse der Willensthähigkeit. *Philosophische Studien, 2*, 73–127.

Miller, G. A. (1956). The magical number seven, plus or minus two: Some limits on our capacity for processing information. *Psychological Review, 63*, 81–97.

Miller, L. T., & Vernon, P. A. (1992). The general factor in short-term memory, intelligence, and reaction time. *Intelligence, 16*, 5–29.

Miller, L. T., & Vernon, P. A. (1996). Intelligence, reaction time, and working memory in 4- to 6-year-old children. *Intelligence, 22*, 155–190.

Myerson, J., Zheng, Y., Hale, S., Jenkins, L., & Widaman, K. (2003). Difference engines: Mathematical models of diversity in speeded cognitive performance. *Psychonomic Bulletin & Review, 10*, 262–288.

Neisser, U. (1967). *Cognitive psychology*. New York: Appleton-Century-Crofts.

Nettelbeck, T. (1980). Factors affecting reaction time: Mental retardation, brain damage, and other psychopathologies. In: A. T. Welford (Ed.), *Reaction times*, pp. 355–401. New York: Academic Press.

Nettelbeck, T. (1987). Inspection time and intelligence. In: P. A. Vernon (Ed.), *Speed of information processing and intelligence*. Norwood, NJ: Ablex.

Nettelbeck, T. (2003). Inspection time and g. In: H. Nyborg (Ed.), *The scientific study of general intelligence: Tribute to Arthur R. Jensen*, pp. 77–91. New York: Pergamon.

Nettelbeck, T., & Brewer, N. (1976). Effects of stimulus–response variables on the choice reaction time of mildly retarded adults. *American Journal of Mental Deficiency, 81*, 85–92.

Nettelbeck, T., & Lally, M. (1976). Inspection time and measured intelligence. *British Journal of Psyhology, 67*, 17–22.

Nettelbeck, T., & Rabbitt, P. M. A. (1992). Aging, cognitive performance, and mental speed. *Intelligence, 16*, 189–205.

Nettelbeck, T., & Wilson, C. (1985). A cross-sequential analysis of developmental differences in speed of information processing. *Journal of Experimental Child Psychology, 40*, 1–22.

Neubauer, A. C. (1991). Intelligence and RT: A modified Hick paradigm and a new RT paradigm. *Intelligence, 15*, 175–192.

Neubauer, A. C. (1997). The mental speed approach to the assessment of intelligence. In: J. S. Carlson, J. Kingma, & W. Tomic (Eds), *Advances in cognition and educational practice: Reflections on the concept of intelligence*, Vol. 4, pp. 149–173. Greenwich, CT: JAI Press, Inc.

Neubauer, A. C., & Bucik, V. (1996). The mental speed–IQ relationship: Unitary or modular? *Intelligence, 22*, 23–48.

Neubauer, A. C., & Fruedenthaler, H. H. (1994). Reaction times in a sentence-picture verification test and intelligence; individual strategies and effects of extended practice. *Intelligence, 19*, 193–218.

Neubauer, A. C., & Knorr, E. Y. (1998). Three paper-and-pencil tests for speed of information processing: Psychometric properties and correlations with intelligence. *Intelligence, 26*, 123–151.

Neubauer, A. C., Spinath, F. M., Riemann, R., Borkenau, P., & Angleitner, A. (2000). Genetic and environmental influences on two measures of speed of information processing and their relation to psychometric intelligence: Evidence from the German observational study of adult twins. *Intelligence, 28*, 267–289.

Niemi, P., & Näätänen, R. (1981). Foreperiod and simple reaction time. *Psychological Bulletin, 89*, 133–162.

Oberauer, K. Schulzem, R., Wilhelm, O., & Süß, H.-M. (2005). Working memory and intelligence their correlation and their relation: Comment on Ackerman, Beier and Boyle (2005). *Psychological Bulletin, 131*, 61–65.

O'Connor, T. A., & Burns, N. R. (2003). Inspection time and general speed of processing. *Personality and Indicidual Differences, 35*, 713–724.

Okada, R. (1971). Decisions latencies in short-term recognition memory. *Journal of Experimental Psychology, 90*, 27–32.

Pachella, R. G. (1974). The interpretation of reaction time in information processing research. In: B. H. Kantorowitz (Ed.), *Human information processing: Tutorials in performance and cognition.* Hillsdale, NJ: Erlbaum.

Pashler, H. (1993). Doing two things at the same time. *American Scientist, 81*, 48–55.

Paul, S. M. (1984). *Speed of information processing: The semantic verification test and general mental ability.* Unpublished doctoral dissertation. University of California, Berkeley.

Peak, H., & Boring, E. G. (1926). The factor of speed in intelligence. *Journal of Experimental Psychology, 9*, 71–94.

Pellegrino, J. W., & Kail, R., Jr. (1982). Process analyses of spatial aptitude. In: R. J. Sternberg (Ed.), *Advances in the psychology of human intelligence*, Vol. 1. Hillsdale, NJ: Erlbaum.

Petrill, S. A., Thompson, L. A., & Detterman, D. K. (1995). The genetic and environmental variance underlying elementary cognitive tasks. *Behavioral Genetics, 25*, 199–209.

Phillips, L. H., & Rabbitt, P. M. A. (1995). Impulsivity and speed-accuracy strategies in intelligence test performance. *Intelligence, 21*, 13–29.

Plomin, R., DeFries, I. C., & McClearn, G. E. (1990). *Behavioral Genetics: a Primer* (2nd ed.) New York: W. H. Freeman.

Poppel, E. (1994). Temporal mechanisms in perception. *International Review of Neurobiology, 37*, 185–202.

Posner, M. I. (1978). *Chronometric explorations of mind.* Hillsdale, NJ: Erlbaum.

Posner, M., Boies, S., Eichelman, W., & Taylor, R. (1969). Retention of visual and name codes of single letters. *Journal of Experimental Psychology, 81*, 10–15.

Rabbitt, P. (1979). How old and young subjects monitor and control responses for accuracy and speed. *British Journal of Psychology, 70*, 305–311.

Rabbitt, P. (1996). Do individual differences in speed reflect "global" or "local" differences in mental abilities? *Intelligence, 22*, 69–88.

Rabbitt, P. (2002). Aging and cognition. In: H. Pashler, & J. Wixted (Eds), *Stevens' handbook of experimental psychology*, 3rd ed., Vol. 4, pp. 793–860. New York: Wiley.

Raz, N., Willerman, L., Ingmundson, P., & Hanlomnf, M. (1993). Aptitude-related differences in auditory recognition masking. *Intelligence, 7*, 71–90.

Rijsdijk, F. V., Vernon, P. A., & Boomsma, D. I. (1998). The genetic basis of the relation between speed-of-information-processing and IQ. *Behavioral Brain Research, 95,* 77–84.

Rink, J. (Ed.) (1995). *The practice of performance: Studies in musical interpretation.* Cambridge, UK: Cambridge University Press.

Roberts, R. D., & Pallier, G. (2001). Individual differences in performance on elementary cognitive tasks (ECTs): Lawful vs. Problematic parameters. *Journal of General Psychology, 128*(3), 279–314.

Roberts, R. D., & Stankov, L. (1999). Individual differences in speed of mental processing and human cognitive abilities: Toward a taxonomic model. *Learning and Individual Differences, 11,* 1–120.

Robinson, M. D., & Tamir, M. (2005). Neuroticism as mental noise: A relation between neuroticism and reaction time standard deviations. *Journal of Personality and Social Psychology, 89,* 107–114.

Roth, E. (1964). Die Geschwindigkeit der Verarbeitung von Information und ihr Zusammenhang mit Intelligenz. *Zeitschrift für Experimentelle under Angewandte Psychologie, 11,* 616–622.

Rowe, F. B. (1983). Whatever became of poor Kinnebrook? *American Psychologist, 38,* 851–852.

Saccuzzo, D. P., Larson, G. E., & Rimland, B. (1986). *Speed of information processing and individual differences in intelligence.* Research Report NPRDC TR 86-23. San Diego, CA: Navy Personnel Research and Development Center.

Salthouse, T. A. (1991). *Theoretical perspectives on cognitive aging.* Hillsdale, NJ: Erlbaum.

Salthouse, T. A. (1993). Attentional blocks are not responsible for age-related slowing, *Journal of Gerontology, 48,* 263–270.

Salthouse, T. A. (1994). Processing speed as a mental capacity. *Acta Psychologica, 86,* 199–225.

Salthouse, T. A. (1996). The processing-speed theory of adult age differences in cognition. *Psychological Review, 103,* 403–428.

Salthouse, T. A. (1998). Relation of successive percentiles of reaction time distributions to cognitive variables and adult age. *Intelligence, 26,* 153–166.

Salthouse, T. A., Fristoe, N. M., Lineweaver, T. T., & Coon, V. E. (1995). Aging of attention: Does the ability to divide decline? *Memory and Cognition, 23,* 59–71.

Schweizer, K. (1998). Complexity of information processing and the speed–ability relationship. *Journal of General Psychology, 125,* 89–102.

Sesardic, N. (2005). *Making sense of heritability.* Cambridge: Cambridge University Press.

Sharp, S. E. (1898–1899). Individual psychology: A study in psychological method. *American Journal of Psychology, 10,* 329–391.

Shepard, R. N., & Metzler, J. (1971). Mental rotation of three-dimensional objects. *Science, 171,* 701–703.

Simmons, R. W., Wass, T., Thomas, J. D., & Riley, P. (2002). Fractionated simple and choice reaction time in children with prenatal exposure to alcohol. *Alcoholic Clinical and Experimental Research, 26,* 1412–1419.

Simonton, D. K. (2003). Genius and *g*: Intelligence and exceptional achievement. In: H. Nyborg (Ed.), *The scientific study of general intelligence: Tribute to Arthur R. Jensen,* pp. 229–246. New York: Pergamon.

Smith, G., & Stanley, G. (1983). Clocking *g*: Relating intelligence and measures of timed performance. *Intelligence, 7,* 353–368.

Smith, G. A., & Brewer, N. (1985). Age and individual differences in correct and error reaction times. *British Journal of Psychology, 76,* 199–203.

Smith, G. A., & Bewer, N. (1995). Slowness and age: Speed–accuracy mechanisms. *Psychology and Aging, 10,* 238–247.

Smith, G. A., & Carew, M. (1987). Decision time unmasked: Individuals adopt different strategies. *Australian Journal of Psychology, 39,* 339–352.

Smith, G. A., & McPhee, K. A. (1987). Performance on a coincidence timing task correlates with intelligence. *Intelligence, 11,* 161–167.

Smith, G. A., Poon, L. W., Hale, S., & Myserson, J. (1988). A regular relationship between old and young adults' latencies on their best, average, and worst trials. *Australian Journal of Psychology, 40,* 195–210.

Smyth, M., Anderson, M., & Hammond, G. (1999). The modified blink reflex and individual differences in speed of processing. *Intelligence, 27,* 13–35.

Sokal, M. M. (Ed.) (1981). *An education in psychology: James McKeen Cattell's journal and letters from Germany and England, 1880–1888.* Cambridge, MA: MIT Press.

Spearman, C. E. (1904). "General intelligence," objectively determined and measured. *American Journal of Psychology, 15,* 201–293.

Spearman, C. (1927). *The abilities of man: Their nature and measurement.* New York: Macmillan.

Sternberg, S. (1966). High speed scanning in human memory. *Science, 153,* 652–654.

Sternberg, S. (1969). Memory scanning: Mental processes revealed by reaction-time experiments. *American Scientist, 57,* 421–457.

Sternberg, R. J. (1977). *Intelligence, information processing, analogical reasoning: The componential analysis of human abilities.* Hillsdale, NJ: Erlbaum.

Stevens, S. S. (1975). *Psychophysics: Introduction to its perceptual, neural and social aspects.* New York: Wiley.

Stokes, T. L., & Bors, D. A. (2001). *Intelligence, 29,* 247–261.

Stough, C., & Bates, T. C. (2004). The external validity of inspection time as a measure of intelligence. Unpublished Report.

Strogatz, S. (2003). *Sync: The emerging science of spontaneous order.* New York: Hyperion.

Stroop, J. R. (1935). Studies of interference in serial verbal reactions. *Journal of Experimental Psychology, 18,* 643–662.

Teichner, W. H., & Krebs, M. J. (1974). Laws of visual choice reaction time. *Psychological Review, 81,* 75–98.

Thorndike, E. L., Bregmann, E. O., Cobb, M. V., & Woodyard, E. (1927/1973). *The measurement of intelligence.* New York: Teachers College, Columbia University/Arno Press.

Thorndike, E. L., & Lorge, I. (1944). *The Teacher's Word Book of 30,000 Words.* New York: Teachers College, Columbia University.

Verhaeghen, P., & Salthouse, T. A. (1997). Meta-analyses of age-cognition relations in adulthood: Estimates of linear and nonlinear age effects and structural models. *Psychological Bulletin, 122,* 231–249.

Vernon, P. A. (1981a). *Speed of information processing and general intelligence.* Unpublished doctoral dissertation. University of California, Berkeley.

Vernon, P. A. (1981b). Reaction time and intelligence in the mentally retarded. *Intelligence, 5,* 345–355.

Vernon, P. A. (1983). Speed of information-processing and general intelligence. *Intelligence, 7,* 53–70.

Vernon, P. A. (1987). *Speed of information processing and intelligence.* Norwood, NJ: Ablex.

Vernon, P. A. (1988). Intelligence and speed of information processing. *Human Intelligence Newsletter, 9,* 8–9.

Vernon, P. A. (1989a). The generality of *g. Personality and Individual Differences, 10,* 803–804.

Vernon, P. A. (1989b). The heritability of measures of speed of information-processing. *Personality and Individual Differences, 10,* 573–576.

Vernon, P. A., & Jensen, A. R. (1984). Individual and group differences in intelligence and speed of information processing. *Personality and Individual Differences, 5,* 411–423.

Vernon, P. A., & Kantor, L. (1986). Reaction time correlations with intelligence test scores obtained under either timed or untimed conditions. *Intelligence, 10*, 315–330.

Vernon, P. A., Nador, S., & Kantor, L. (1985). Group differences in intelligence and speed information processing. *Personality and Individual Differences, 10*, 573–576.

Vernon, P. A., & Vollick, D. N. (1998). Speed of information-processing in the elderly: Evidence against a general deficit. Paper presented at the annual meeting of the AERA, New Orleans, USA.

Vickers, D. (1995). The frequency accrual speed test (FAST): a new measure of 'mental speed'? *Personality and Individual Differences, 19*, 863–879.

Vickers, D., Nettelbeck, T., & Willson, R. (1972). Perceptual indices of performance: The measurement of Ainspection time@ and Anoise@ in the visual system. *Perception, 1*, 263–295.

Von Neumann, J. (1941). Distribution of the ration of the mean square successive difference to the variance. *Annals of Mathematical Statistics, 14*, 378–388.

Wechsler, D. (1944). *The measurement of intelligence* (3rd ed.). Baltimore: Williams & Wilkins.

Weiss, D. J. (1995). Improving individual differences measurement with item response theory and computerized adaptive testing. In: D. Lubinski, & R. Dawis (Eds), *Assessing individual differences in human behavior: New methods, concepts, and finding*, pp. 49–79. Palo-Alto, CA: Davies-Black Publishing.

Welford, A. T. (Ed.) (1980a). *Reaction times*. London: Academic Press.

Welford, A. T. (1980b). Choice reaction time: Basic concepts. In: A. T. Welford (Ed.), *Reaction times*, pp. 73–128. London: Academic Press.

Welford, A. T. (1980c). Relationships between reaction time and fatigue, stress, age and sex. In: A. T. Welford (Ed.), *Reaction times*, pp. 321–349. New York: Academic Press.

Wickens, C. D. (1974). Temporal limits of human information processing: A developmental study. *Psychological Bulletin, 8*, 739–755.

Widaman, K. F., & Carlson, J. S. (1989). Procedural effects on performance on the Hick paradigm. *Intelligence, 13*, 63–86.

Wilson, C., & Nettelbeck, T. (1986). Inspection time and the mental age deviation hypothesis. *Personality and Individual Differences, 7*, 669–675.

Wissler, C. (1901). The correlation of mental and physical traits. *Psychological Monographs, 3*, 1–62.

Woodworth, R. S. (1938). *Experimental psychology*. New York: Holt.

Woodworth, R. S., & Schlosberg, H. (1956). *Experimental psychology*. New York: Holt.

Zhang, Y. (1991). *Personality and individual differences, 12*, 217–219.

# Jensen References on Chronometry Not Cited in the Text

Buckhalt, J., & Jensen, A. R. (1989). The *British Ability Scales* speed of information processing subtest: What does it measure? *British Journal of Educational Psychology, 59*, 100–107.

Jensen, A. R. (1980). Chronometric analysis of intelligence. *Journal of Social and Biological Structures, 3*, 103–122.

Jensen, A. R. (1981). Reaction time and intelligence. In: M. Friedman, J. P. Das, & N. O. Connor (Eds), *Intelligence and learning,* pp. 39–50. New York: Plenum.

Jensen, A. R. (1982). The chronometry of intelligence. In: R. J. Sternberg (Ed.), *Advances in the psychology of human intelligence*, Vol. 1, pp. 255–310. Hillsdale, NJ: Erlbaum.

Jensen, A. R. (1984). Mental speed and levels of analysis. *The Behavioral and Brain Sciences, 7*, 295–296.

Jensen, A. R. (1987). The *g* beyond factor analysis. In R. R. Ronning, J. A. Glover, J. C. Conoley, & J. C. Witt (Eds), *The influence of cognitive psychology on testing,* pp. 87–142. Hillsdale, NJ: Erlbaum.

Jensen, A. R. (1987). Mental chronometry in the study of learning disabilities. *Mental Retardation and Learning Disability Bulletin, 15*, 67–88.

Jensen, A. R. (1988). Speed of information processing and population differences. In: S. H. Irvine (Ed.), *The cultural context of human ability.* London: Cambridge University Press.

Jensen, A. R. (1992). Understanding *g* in terms of information processing. *Educational Psychology Review, 4*, 271–308.

Jensen, A. R. (1992). The relation between information processing time and right/wrong responses. *American Journal on Mental Retardation, 97*, 290–292.

Jensen, A. R. (1993). Psychometric *g* and achievement. In: B. R. Gifford (Ed.), *Policy perspectives on educational testing,* pp. 117–227. Norwell, MA: Kluwer Academic Publishers.

Jensen, A. R. (1993). Test validity: *g* versus "tacit knowledge." *Current Directions in Psychological Science, 2*, 9–10.

Jensen, A. R. (1993). Why is reaction time correlated with psychometric *g*? *Current Directions in Psychological Science, 2*, 53–56.

Jensen, A. R. (1993). Spearman's hypothesis tested with chronometric information processing tasks. *Intelligence, 17*, 47–77.

Jensen, A. R. (1994). Reaction time. In: R. J. Corsini (Ed.), *Encyclopedia of psychology*, 2nd ed., Vol. 3, pp. 282–285. New York: Wiley.

Jensen, A. R. (1996). Inspection Time and *g*. (Letter), *Nature, 381*, 729.

Jensen, A. R. (2000). Elementary cognitive tasks and psychometric g. In: A. Harris (Ed.), *Encyclopedia of psychology*, Vol. 3, pp. 156–157. New York: APA/Oxford University Press.

Jensen, A. R., & Reed, T. E. (1992). The correlation between reaction time and the ponderal index. *Perceptual and Motor Skills, 75,* 843–846.

Jensen, A. R., & Whang, P. A. (1993). Reaction times and intelligence: A comparison of Chinese-American and Anglo-American children. *Journal of Biosocial Science, 25,* 397–410.

Jensen, A. R., & Whang, P. A. (1994). Speed of accessing arithmetic facts in long-term memory: A comparison of Chinese-American and Anglo-American children. *Contemporary Educational Psychology, 19,* 1–12.

Kranzler, J. H., & Jensen, A. R. (1989). Inspection time and intelligence: A meta-analysis. *Intelligence, 13,* 329–347

Reed, T. E., & Jensen, A. R. (1991). Arm nerve conduction velocity (NCV), brain NCV, reaction time, and intelligence. *Intelligence, 15,* 33–47.

Reed, T. E., & Jensen, A. R. (1993). Choice reaction time and visual pathway nerve conduction velocity both correlate with intelligence but appear not to correlate with each other: Implications for information processing. *Intelligence, 17,* 191–203.

Sen, A., Jensen, A. R., Sen, A. K., & Arora, I. (1983). Correlation between reaction time and intelligence in psychometrically similar groups in America and India. *Applied Research in Mental Retardation, 4,* 139–152.

# Author Index

# Subject Index